# Research with Recombinant DNA

AN ACADEMY FORUM
March 7–9, 1977

NATIONAL ACADEMY OF SCIENCES
Washington, D.C.     1977

The work upon which this publication is based was performed pursuant to
support from the National Institutes of Health, Contract Number NOI-OD-
7-2104.


Library of Congress Catalog Card Number 77-86491

International Standard Book Number 0-309-02641-5


*Available from:*

Printing and Publishing Office
National Academy of Sciences
2101 Constitution Avenue, N.W.
Washington, D.C.   20418

Printed in the United States of America

# CONTENTS

DAY II

---

DAY III

# FOREWORD

This publication has been developed from the proceedings of the Forum on "Research with Recombinant DNA," convened at the National Academy of Sciences from March 7-9, 1977.  It contains a variety of insights and information concerning the research, as well as the controversy that has encompassed it.

The planners and implementers of this Academy Forum wish to acknowledge support received from:  Abbott Laboratories, Ford Foundation, Hoffmann-La Roche, Inc., Eli Lilly and Company, Merck Sharp and Dohme Research Laboratories, Miles Laboratories, the National Academy of Sciences, the National Institutes of Health, Pfizer Inc., Smith Kline & French Laboratories, and the Upjohn Company.

Although it is impossible to cite all of those who helped this Academy Forum become a reality, some special acknowledgment should be made to: David Hamburg and Alexander Rich, who piloted the proceedings through many hours with poise, patience, and judgment; the General Advisory Committee of the Academy Forum, particularly its chairman Daniel E. Koshland, Jr., for selecting this topic and encouraging it through a complex development; M. Virginia Davis for the astute research and planning she gave to this Forum; all those members of the National Academy of Sciences staff who came forward with logistical and moral support as attendance passed all records in the heat of dissension and television lights.

Many decisions remain to be made about the future of research with recombinant DNA.  We hope that this Forum and its publication will be of some help to those who ultimately must make them and to those who must accept them.

Robert R. White
*Director*

# INTRODUCTION TO THE FORUM

Daniel E. Koshland, Jr.

*Chairman, General Advisory Committee of the Academy Forum; Professor and Chairman, Department of Biochemistry, University of California, Berkeley*

The Academy Forum was initiated by the National Academy of Sciences to make a contribution to national policy in areas at the interface of science and society. Past Forums have dealt with such controversial issues as drug safety, experimentation with humans, utilization of energy resources, and the roles of the citizen and the expert. The conferences and the publications resulting from these Forums have had widespread use. None of the past Forums, however, has dealt with a subject at such a height of controversy or so poised at the brink of incipient local, national, and international legislation as this one. Its participants, therefore, had a signal chance to make a contribution to the design of public policy, and I believe they have done so. This Forum had many aspects, but two themes are illustrative of the difficulties of designing science policy. The first of these is the classic confrontation between the innovator and regulator. The second is the difficulty of communication between the scientist and the layman. The issues are particular to this Forum, but the fundamentals lie in the background of all attempts to reach a consensus on science policy.

It is quite clear that a society must be based on some mutually acceptable moral standards. But morality is frequently a poor basis for decision making since each protagonist is convinced that virtue stands resolutely at his side. Speaking oversimply, the job of the scientist is the creation of goodness; the job of the regulator is the prevention of evil. As knowledge has become more complex, the scientist with his laboratories and computers becomes modern society's explorer of new horizons. Like all adventurers he feels his tasks are arduous enough without the burdens of red tape and ponderous legalities. Regulators, on the other hand, remember the history of the swashbuckling buccaneer and the robber baron and see their job as protecting society from the unbridled ambitions of a few. Scientists do not wish to do harm, and regulators do not wish to stifle progress, and yet their differing needs and desires inevitably make the scientist hostile to control and the regulator conservative about progress. Both the creation of goodness and the prevention of evil are worthy and moral goals, and a choice therefore requires one to establish a fine line between the benefits and risks to society.

In trying to establish the benefits and risks of recombinant DNA techniques, the problem of communication between the scientist and the layman inevitably becomes difficult: this is illustrated poignantly in this Forum where the scientific community itself is far from monolithic. Since the scientific community is split, a great deal of the discussion involved the attempt of one group of scientists to convince another group, and in the process they inevitably slipped into the jargon of science. Words such as *DNA, plasmids, virus, vectors,* etc., are utilized not to exclude a lay public, but because they represent a shorthand notation of accumulated knowledge and theory that allows for rapid and precise communication. To help bridge this gap somewhat, John Abelson has written a brief introduction to the science of recombinant DNA defining terms and concepts. And many of the scientific presentations have been carefully phrased in nontechnical words to provide maximum information to the nonscientist. It is important to remember, however, that we are dealing with a subject on the frontiers of science that requires some technical language. Most of us are used to arguing with our peers: scientist with scientist, ethicist with ethicist, but bridging the language gap is almost as difficult as bridging the philosophical gap. We need to develop social mechanisms, of which we believe the Academy Forum is one, to help close the interdisciplinary gap and create this common understanding.

To aid in this mechanism this Forum initiated two new techniques, the case analysis approach and the workshops. The case analysis attempted to focus the discussion toward more concrete problems by taking specific examples: the basic research benefit-risk, the insulin production benefit-risk, etc. The workshops provided a mechanism for extended discussion in a specific area.

In the Academy Forum process the publication of a volume occupies a crucial role. A book provides a variety of benefits. It multiplies the audience, it rewards economy of language, and it provides a constraint on extravagant claims. The Academy Forum itself provides an initial constraint by the distinction and expertise of the audience, but the book is vital. At a recent symposium one speaker paused in mid-sentence and looked in horror at the taping machine. "Are my comments being recorded?" he asked. The chairman said "Yes, but why do you care?" The reply was "I don't mind making a fool of myself in public but I hate to do it in print." Yesterday's newspapers are thrown away, but the scientific volume remains for a long time. Academy Forum volumes are constantly being used even years after their initial publication. We have found that the knowledge that remarks are to be recorded for history has great therapeutic value in keeping remarks to those which can be justified, thus expediting the approach to a final consensus.

Unlike some other areas of public policy, science tends to glorify the sacredness of a fact. Some controversy simply involves differences in moral judgments, and then instinctive or hereditary cultural values control opinions. In the case of science policy, value judgments enter but the facts are crucial. The instinct of the scientist is to delay decisions until all the facts have been accumulated. Unfortunately, this is not practical in areas such as nuclear reactors, red dye #2, or recombinant DNA. The scientist inevitably pleads for more time to discover his facts. The layman wishes to push the legislation "before the damage is done." Action does not, however, lessen the importance of separating fact from fiction. In the current dispute, some individuals refer to the "Andromeda strain" as though it had been created in the laboratory and was not merely a work of fiction. Failure to distinguish fact from fiction can lead to fundamental disagreement even among those who believe that both safety and progress are desirable goals.

In this regard I might mention a story about a Berkeley colleague who died recently. As his heirs were going over his estate, they found a set of aged lecture notes which contained in the margins exhortatory instructions. One such notation said, "Speak loudly here, uncertain of fact." There is a tendency in controversial legislative issues to believe that loud speaking or flamboyant publicity is a substitute for hard facts. In some vital public issues, such as labor disputes or territorial boundaries, a middle compromise is frequently appropriate. In science, however, if one group says that 2 + 2 is 4, and another that 2 + 2 is 5, it does not follow that 2 + 2 is 4½. This volume, therefore, is an attempt to help the layman and the scientist understand those facts that are already established, those facts that are in dispute, and the practical and moral considerations that must be part of any attempt to make final policy.

A volume of proceedings therefore allows the reader to examine at leisure (and with the additional facts developed over time) the arguments of the protagonists. Hopefully, this will allow a wider constituency to inform themselves and clarify the issues in this vital area of scientific development.

# INTRODUCTION TO
# RECOMBINANT DNA RESEARCH

John Abelson

*Professor of Chemistry*
*The University of California, San Diego*

The papers in this volume explore and discuss the scientific, legal, and moral issues that have been raised by the recombinant DNA technology. Because this subject has become one of importance not only for scientists but the public as well, it is desirable that there be some general understanding of the science involved. Fortunately it is not difficult to attain an elementary knowledge of the fundamental principles in recombinant DNA research. I will attempt here to provide a primer that is intended to enable interested persons to better understand the contents of this volume.[1]

Although a superficial survey of life on this planet reveals a bewildering diversity of living forms, research into the chemistry of life reveals that, in fact, there is an amazing unity. Organisms utilize only a small fraction of the possible compounds that can be formed by the elements carbon, nitrogen, hydrogen, oxygen, phosphorous, and sulfur (the principle elements found in most organisms). By and large the same compounds are found in all organisms from bacteria to man.

This unity is perhaps most striking in the case of the molecular basis of heredity. It is obvious that there must be a stable repository of information that is passed on from one generation of a species to the next. In certain species (e.g., some clams) this information must have been preserved in essentially unchanged form for hundreds of millions of years. We now know that genetic information is encoded in the macro-molecule deoxyribonucleic acid (DNA) and that this is the carrier of genetic information in all species (with the exception of some viruses where it is the closely related molecule RNA).

Probably one of the most exciting moments in science came when it was realized by James Watson and Francis Crick in 1953 that the *structure* of DNA could explain its *function* as the carrier of genetic information.

DNA is a polymeric molecule made up of four different monomer units. The polymeric molecule is always a linear (or circular) chain of monomer units. The use of the words *monomer* and *polymer* in this case is analogous to beads and strings of beads. The monomer units are called nucleotides (so polymers of these units are sometimes called polynucleotides). Each nucleotide unit is composed of a base (there are four different bases: adenine [A], guanine [G], thymine [T], or cytosine [C], a sugar (deoxyribose), and a phosphate group (PO$_4$). A phosphate group between two adjacent sugars in the chain links the nucleotides to form polynucleotides (Figure 1).

Watson and Crick discovered that DNA is composed of two polynucleotide chains intertwined about each other to form a double helix. The two chains are paired together via weak forces between the bases. A base in one chain always pairs with a base in the opposite chain according to the base-pairing rules (Figure 2). These rules are implicit in the chemical properties of the four bases, such that A always pairs with T and G always pairs with C. Thus the sequence of the bases in one strand of the double helix uniquely determines the sequence of bases in the opposite strand (Figure 3).

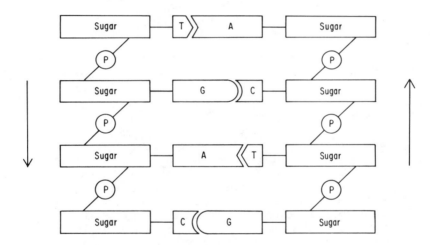

FIGURE 1  A schematic diagram of the organization of nucleotides in DNA molecules. This is a diagram of a small section of the double helix. The backbone of the polynucleotide chain consists of alternating sugar and phosphate (P) molecules. The two chains are held together by weak bonds between the bases (see Figure 2). Note that the chains have a direction which is determined by the sugar and that they are antiparallel. In a short-hand notation that is frequently used the left-hand strand would be referred to as TpGpApC and the right-hand strand as GpTpCpA.

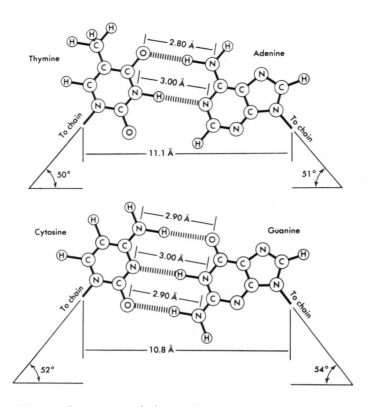

FIGURE 2 Base pairing. The chemical structures
of the four bases are shown, as are the interactions between them. The weak forces between the
bases are called hydrogen bonds because a hydrogen atom is partially shared between two bases,
e.g., the hydrogen atom that is part of the
adenine molecule is attracted to the oxygen atom
of a thymine molecule. The structure of these
base pairs is such that both pairs have the same
shape and therefore both fit into the double
helix in the same way. It can easily be seen
why adenine-guanine or cystosine-thymine pairs
would not work. (From J. D. Watson, *The Molecular Biology of the Gene*.)

    Watson and Crick immediately realized that the structure of DNA could
explain its unique role as the carrier of genetic information. First
one must postulate that the genetic information is encoded in the
sequence of bases in the polynucleotide chain. This postulate is now
supported by so much experimental evidence that it can be considered
as fact. It is at this point that *chemistry* and *genetics* fuse.

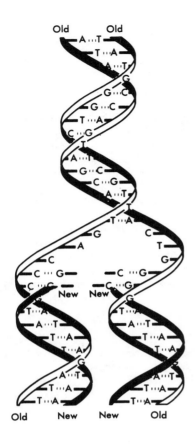

FIGURE 3  The double helical structure of DNA.  This diagram shows how the two poly-nucleotide chains intertwine to form a double helix.  The diagram also illustrates how a parent DNA molecule could be repli-cated to form two identical daughter mole-cules.  The two strands of the double helix would separate, and each would serve as the template for the synthesis of a complemen-tary strand.  Individual nucleotides are incorporated at the correct position ac-cording to the base pairing rules.  (From J. D. Watson, *The Molecular Biology of the Gene.*)

Starting with Mendel's discoveries in the nineteenth century, geneticists came to realize that the heritable properties of organisms are organized as genes, and by 1945 George Beadle realized that each gene must control the structure of a protein.  We now know that genes are segments of DNA and that encoded in the sequence of bases in each gene is the structure of a protein.  There is a simple hierarchy in the cell.  The information for the structure of proteins is carried in the sequences of bases in the DNA.  The proteins in turn determine the structure and chemical capabilities of the organism.  Proteins can serve a structural role, for example, in muscle or connective tissue or as enzymes they can catalyze the chemical reactions that enable the organism to extract energy from its environment and utilize that energy for growth and reproduction.

Proteins also are linear polymers composed not of four but of twenty different monomers-amino acids.  The sequence of the amino acids in a protein determines its properties.  This sequence is dictated by the sequence of bases in the gene.  A set of three contiguous bases *code* for an amino acid.  When the four bases are divided into groups, three at a time, there are sixty-four triplets or codons. The assignment of these codons to amino acids is called the genetic code. There is now overwhelming evidence that the code is universal--that it is the same in all organisms.

The double helix also explains how copies of the genetic information could be passed on to succeeding generations. One need only postulate that the strands of the double helix separate and that each strand serves as a *template* for the synthesis of a new complementary strand (see Figure 3). The base-pairing rules ensure that two identical copies of the parental DNA will result. Although the process of DNA synthesis is complicated, it is now known that DNA replication does proceed essentially in this way.

To summarize, the structure of DNA explains its function as the repository of genetic information. The sequence of bases in each gene, a segment of DNA approximately 1,000 base pairs in length, determines the structure of a protein. Because the code is universal there is at least a potential principle of compatibility between all organisms. The gene from a tomato which encodes the information for a particular protein would encode information for synthesis of the same protein if that gene were present in a fish.[2]

It should be mentioned at this point that a copy of the hereditary information (DNA) of an organism is contained in each cell whether the organism be a unicellular bacterium or a complicated multicellular mammal. The amount of DNA per cell is roughly proportional to the complexity of the organism. Thus bacteria contain enough DNA to encode the information for about 3,000 proteins; *Drosophila,* a fruit fly, contains about 30 times as much DNA per cell; and mammals, about 1,000 times as much. The electron micrographs in Figures 4 and 5 provide a graphic display of the relative complexities of the bacterial virus T2, and the bacterium *Hemophilus influenzae.* T2 contains about 100 genes, and *Hemophilus,* about 3,000 genes. The length of the DNA in a single human cell would be at least 1,000 times the length of the *Hemophilus* DNA.

We are now prepared to discuss the actual details of the recombinant DNA technology.[3] *Recombination* as used here is a genetic term. If, for example, two strains of mutant bacteria--one unable to grow without added tryptophan (an amino acid) and the other unable to ferment lactose (a sugar)--are mated, a small percentage of the progeny will be "wild type," that is, able to grow without tryptophan and able to ferment lactose. At the molecular level, this recombination between the two strains results because of breakage of the parental DNA molecules and subsequent rejoining to form the recombinants.

The recombinant DNA technology has made it possible to carry out this recombination process of breakage and rejoining of DNA molecules in the test tube. For many years geneticists used the recombination process to study the nature of genes, but they were confined to studying recombinants of the same or closely related species. The *in vitro* process now makes it possible, for the first time, to join DNA molecules of unrelated organisms.

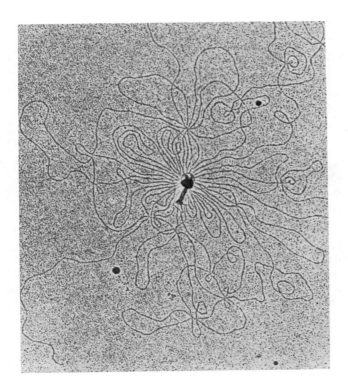

FIGURE 4 Electron micro-
graph of T2 bacteriophage
DNA. The magnification
in this micrograph is
100,000. The DNA has
been released from the
tadpole-shaped virus
whose remains can be
seen in the center of
the micrograph. (From
A. K. Kleinschmidt *et
al.*, Biochim. Biophys.
Acta 61:857, 1962.)

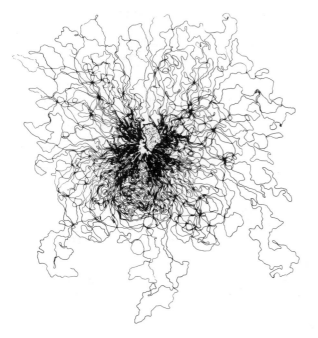

FIGURE 5 Tracing of an
electron micrograph of *Hemophilus influenzae* DNA. The
magnification is about
12,000. Again the DNA is
released from the bacterium,
whose remains are seen in
the center. (From L. A.
MacHattie, K. I. Berns, and
C. A. Thomas, Jr., J. Mol.
Biol. 11:648, 1965.)

The present capability to join DNA molecules *in vitro* is the result of twenty-five years of research on the properties of DNA and the enzymes involved in cleaving and replicating it. In this process, the most important enzymes are the restriction endonucleases. These enzymes have the capability to cut DNA at sequence-specific sites. In Roberts' recent review of this subject,[4] more than eighty such enzymes isolated from diverse strains of bacteria are described. Some of these enzymes produce staggered cuts in the double strands so that single-stranded ends are produced at each end of the fragment (Figure 6). All fragments produced by a given enzyme have the same self-complementary end, so that a single fragment can circularize by base pairing or it may combine with another fragment to produce a dimer. In the latter case, if the fragments are from different sources, a recombinant molecule is produced. When this technique is used to produce recombinant DNAs,

FIGURE 6 Outline of the procedures involved in construction and propagation of a recombinant DNA molecule. Vehicle and passenger DNA are cleaved with a restriction endonuclease. One (of several) restriction endonuclease that is used for this purpose is obtained from *E. coli* and is called *Eco Rl*. It cleaves the DNA sequence:

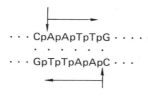

to form

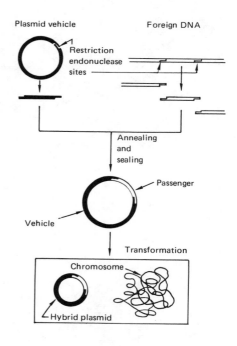

This sequence of six bases will occur by chance about once every 4,000 bases.

The self-complementary sequences are sealed by DNA ligase to form recombinant DNA molecules consisting of both vehicle and passenger. These molecules are introduced into *E. coli* (represented by the rectangle at the bottom) by the process of transformation.

the gaps remaining in both strands are sealed by an enzyme called DNA ligase.

It is convenient to refer to the two DNA molecules in a recombinant as the vehicle and the passenger (Figure 6). The vehicle is a DNA molecule that is capable of self-replication in its host. When the passenger is joined to the vehicle, it behaves as a (sometimes) passive addition and is replicated together with the vehicle. Two general classes of vehicles have been employed in *Escherichia coli*. Plasmids are small, circular DNA molecules that are capable of self-replication independent of the chromosome. Most of the plasmids now in use originated from the plasmid Col El, which is about 1/1,000 the size of the *E. coli* chromosome and produces a bactericidal protein--colicin El-- which kills other bacteria that do not carry the plasmid. The other *E. coli* vehicle is the well-studied bacterial virus, phage $\lambda$. For simplicity we will only discuss the plasmid vehicles.

When the recombination *in vitro* between vehicle and passenger DNA is completed, the recombinant DNA must be introduced into the host cell where it can replicate. This is done by a process known as *transformation*. The DNA is added to *E. coli* cells and is taken up at a low frequency. The plasmid vehicles are constructed such that they give the transformed cell a selective advantage. Some vehicles carry antibiotic resistance genes so that only transformed cells can grow in the presence of the antibiotic. It is the transformation process that allows one to "clone" DNA fragments. Since the frequency of transformation is low, a transformed cell is, in general, produced by the uptake of a single vehicle-passenger DNA molecule. The transformed cell multiplies to form a colony. Different colonies contain different passengers. Each colony can be propagated indefinitely, enabling one to store or produce (or both) the passenger DNA molecule at will.[5]

In some cases the recombinant DNA approach is used to amplify a passenger DNA molecule that can be purified by other methods. In other cases the passenger DNA joined to the plasmid vehicle is a complex mixture. If the DNA were, for example, from *Drosophila*, there could be about 40,000 different fragments following restriction endonuclease cleavage. Each clone would contain one of those fragments linked to a plasmid vehicle. We would have to isolate some 100,000 clones in order to have a good chance of cloning each one of the fragments. The cloning of random DNA fragments from a particular source has been called a "shotgun experiment." The collection of clones that results is a colony bank. One can search through that bank to find the gene of interest. When a particular clone from the bank is chosen--for example, from a *Drosophila* bank--that DNA fragment has been purified through the procedure of cloning 40,000-fold. This is more than a quantitative improvement over earlier techniques. It would be very difficult to isolate a *Drosophila* gene by any other method.

These new methods make it possible to study the structure of genes from any organism. In the past it has been possible to study the structure of genes of *prokaryotes*, and we understand a great deal about the control of their expression (see Paul Berg's paper in this volume).

Comparatively little is known about the structure and organization of genes in *eukaryotes*.[6] While the recombinant DNA technology can and is being used to isolate genes from prokaryotes, it is in the study of eukaryotes that its greatest application is expected. As can be seen in this book, venturing into this area is both exciting and controversial.

## NOTES

1. This article is partially derived from my paper in Science (195:159, 1977) entitled "Recombinant DNA: Examples of Present Day Research." For a comprehensive background in this field, I recommend *Molecular Biology of the Gene* (J. D. Watson, ed., W. A. Benjamin, Menlo Park, California).

2. Although we believe the code is universal it may well be that the mechanisms of reading a gene are sufficiently different between unrelated organisms that a tomato gene would in fact not be expressed in a fish.

3. For another popular description of these techniques, see S. Cohen, Sci. Am. 233:4, 1975.

4. R. J. Roberts, Crit. Rev. Biochem. 3:123, 1976.

5. Two popular misconceptions should be cleared up at this point. Cloning as referred to here and in recombinant DNA research in general is totally different from the process developed by Gurdon where the nucleus from a somatic cell containing the complete genetic information of one individual (a frog in Gurdon's experiments) is removed and implanted into an enucleated egg. Gurdon demonstrated that this egg could in some cases develop normally to produce a frog. By this technique one could--in principle--propagate an indefinite number of identical individuals. By contrast, in DNA cloning experiments, fragments of passenger DNA represent a minute fraction of the total genetic information of the organism from which they are derived.

    Clones carrying recombinant plasmids are often referred to as hybrid strains. If passenger DNA from the mouse is joined to a plasmid vehicle and transformed into *E. coli,* the resulting clones would be referred to as *E. coli*-mouse hybrids. The use of the word hybrid suggests to the layman that a significant mouse character is acquired by *E. coli*. This is, of course, not the case. The hybrid *E. coli* is still very much the organism that it was. The mouse information that it has acquired represents about one-millionth of the genetic information of the mouse--a few genes at most--and is only about one-thousandth of the genetic information of *E. coli*.

6. Biologists have divided all organisms into two groups called prokaryotes and eukaryotes. The basic difference between these two groups is that in eukaryotes the DNA in each cell is contained in a defined subcellular structure called the nucleus. The nucleus is surrounded by a thin membrane that separates it from the rest of

the cell.  Prokaryotes do not have a nucleus.  All bacteria are
prokaryotes.  Other unicellular organisms, e.g., yeast, are
eukaryotes, as are all higher organisms.  In evolutionary terms
prokaryotes are more primitive than eukaryotes.  Besides the possession of a nucleus there are a number of other differences between
prokaryotes and eukaryotes.  Thus it can be said that we have a
great deal more in common with yeast than we do with bacteria.

# DAY I

DAVID A. HAMBURG
President, Institute of Medicine
*Cochairman*

# PRIORITIES FOR DAY I

David A. Hamburg

This Academy Forum is an attempt by the scientific community to widen the network of ideas, information, and people involved in considering new directions in science. During the past quarter century we have seen a remarkable acceleration in the life sciences. This work has given us profound insight into the nature of living organisms and, I think it is fair to say truly beautiful visions of the nature of the tiny units that make life possible in all of its diversity and complexity.

One aspect of this research, recombinant DNA, is of intense interest to the scientific community; yet it has also generated deep concerns in that community, among the deepest I can recall in my lifetime. This work has ramifications that go beyond the laboratory, go beyond science as discovery, go beyond science even as a basis for developing useful skills pertinent to human suffering. Serious doubts about the limits of predictability and the limits of wisdom have arisen and must be squarely faced. One useful way to do that and one way that we employ consistently in the Institute of Medicine is to widen the circle of those who have access to all the information and ideas on such a subject, draw into reflective consideration the multiple perspectives needed in the long run to arrive at reasonable understanding. Thus, we have brought together for this Forum a remarkable diversity of competences.

There are some here who are expert in recombinant DNA research, and among these are scientists who have many different views of its potentialities for better and worse. There are some scientists who are not expert in that subject but in related subjects: infectious diseases, clinical medicine, public health. There are some scholars whose interests are not in science per se but in the impact of science on society, or the role of society in shaping scientific directions, for

17

example, scholars in ethics and law.  And there are some people who are concerned in the broadest terms with the welfare of the public. Thus, there is here no party line, no royal road to truth, no assumptions of omniscience.  We meet here to consider, analytically if possible, all the information and ideas we can muster bearing on this important area of inquiry.

We have learned that there are people who have severe criticism of the nature of the program.  In the spirit of openess in which the program was conceived, we will call on Jeremy Rifkin of the Peoples Business Commission to explain his criticism of the program.


## DISCUSSION

RIFKIN:  The last speaker said that there was no party line here at the National Academy of Sciences Forum on recombinant DNA.  I am going to take exception to that.  I was told that I would get about five minutes to speak out of the three days.  Another speaker, Jonathan King, said that I could take his five minutes as well, so I will go about nine minutes out of two-and-a-half days, to try to give a critical analysis of why we think this conference is rigged.

Behind all the sanitized phraseology, the technical jargon, there are a few inescapable facts.  First of all, if you will look in the beginning of your program, there are nine sponsoring organizations. Six of those sponsoring organizations are the very pharmaceutical companies that we exposed for doing previously secret research into this whole area of new forms of life.

If you look in the back of the printed program you will see that over two-thirds of the people on the planning committee for this conference come out of the scientific community.  Yet, even by their own acknowledgment, this issue is of such grave consequence to the entire American people that every constituency should be represented fairly and equally in meetings of this magnitude.  I do not believe that 66 percent of the people putting on this Forum should come from just the scientific community.

Many of you have been to these kinds of gatherings before, and we might as well lay it on the table:  they are legitimizing functions for the national press.  We have a controversy here over DNA.  For three days we will hear debate, some token opposition, at the end of which time the *New York Times* is expected to write an article with the headlines, "National Forum on DNA."  It will point out that there was some minority objection included in the conference but the overwhelming consensus was to go ahead with this research with some small qualifications and perhaps some legislation.

The one interesting thing about this Forum is that we are missing the central issue of why the subject is so important.  We have heard for months both proponents and critics arguing that the real question here is safety. Is it safe or unsafe in the laboratories to conduct this experimentation? Do we need P1 laboratories or P4 laboratories? Do we need NIH voluntary guidelines or do we need involuntary regulations?

My friends, the real issue is not whether the laboratory conditions are safe or unsafe, although obviously there is a problem with potential viruses and bacteria getting out of the laboratory and endangering the health and well-being of millions of people. But that is not the central issue. We could have legislation passed this spring by Congress for safety regulations, and it still would not detract from the central issue we are facing.

The real issue here is the most important one that humankind has ever had to grapple with. You know it, and I know it. With the discovery of recombinant DNA scientists have unlocked the mystery of life itself. It is now only a matter of time--five years, fifteen years, twenty-five years, thirty years--until the biologists, some of whom are in this room, will be able literally, through recombinant DNA research, to create new plants, new strains of animals, and even genetically alter the human being on this earth.

Some scientists in attendance will say this is sensationalism, this is emotionalism. Well, it is because this technology is sensational, and because it hits right to the basic emotional core of life itself. For three generations of Americans weaned on Huxley's *Brave New World* the long-range implications of experimentation in this field are ominous. And the precedents are being set right now with the scientists in attendance here.

You can't hide from the fact any more convincingly than the physicists were able to hide from the knowledge that they had when they split the atom. They knew what that could lead to and what it would lead to. Biologists doing DNA research also know what this technology could lead to and what it will inevitably lead to unless there is public reaction.

I recently reread Huxley's *Brave New World*. Among other things, they talk about how to develop social stability through cloning. You all know what that is, the creation of large numbers of genetically identical individuals whose carefully chosen physical and mental qualities make them ideally suited to perform specialized functions. We know that scientists have cloned frogs, but what about human beings?

Well, I thought, it is one thing for us lay people to exaggerate about cloning. So what I did is I went right to your own experts, Dr. Watson and Dr. Lederberg--both Nobel Prize winners in this field. I wanted to hear what they had to say, what their philosophy was about the long-range implications of the techniques they are experimenting with.

First of all, in an article entitled "Moving Toward the Clonal Man," Watson predicts, "That this technology will proceed in its current fashion. A human being born of clonal reproduction will most likely appear on earth within the next 20 to 50 years or sooner if we actively develop it."

Joshua Lederberg says that it is between two and fifteen years. An exaggeration? These are your proponents talking about the research they are involved in. Lederberg, by the way, goes on to favor cloning. He says, "If a superior individual is identified why not copy it directly rather than suffer all the risks of recombinational disruption including those of sex. Leave sexual reproduction for experimental purposes. When a suitable type"--meaning person--"is ascertained, take careful care to maintain it by clonal propagation."

How about Dr. James Bonner of Cal Tech? He has actually come up with a suggestion on how you can actually apply your research, because we know the scientists here aren't just involved in pure research. One suggestion he makes is to remove genetic material from each individual immediately after birth, and then promptly sterilize the individual. "During the individual's lifetime records would be kept of accomplishments and characteristics. After the individual's death a committee decides if the accomplishments are worthy of procreation into other individuals. If so, genetic materials would be removed from the refrigerator depository, and stimulated to clone a new individual. If the committee decides the genetic material is unworthy of procreation it is destroyed." He says, in summing up, "The question indeed is not a moral one, but a temporal one. When do we begin?" For God's sake, Dr. Bonner, Dr. Lederberg, Dr. Watson, your people, your proponents, the leaders in this field of recombinant DNA, those are their views on the long-range implication of this research.

Now, let me ask the scientists here a question. How many scientists and corporate executives from the pharmaceutical companies in this room believe that they have a moral right and an authority to proceed on this experimental path before the American people, all 200 million, are fully informed about all, good and bad, of the long-range implications of this research? How many scientists here believe that they have the moral authority to continue before the American people know about all the implications, pro and con? Stand up. Let us see. We can count you for the television cameras. Or the people all over the world. One person? Two? That is not bad, and I hope the *New York Times* will record that. Only two people are going to proceed now. We don't even need a moratorium. We have had a voluntary one.

Now, the scientists here in this room talk about going to the public, and I actually heard a press conference this afternoon where they were honestly saying that the public is starting to become well informed on this issue. The American public doesn't know anything about this issue yet. How many people on this committee right here would agree to seven days and seven nights of prime time television, just like "Roots," on all three networks to discuss all the implications, pro and con, of genetic engineering? How many? One.

What do you think the American public would say if Watson, Lederberg, and Bonner, who I just quoted, were to share their long-range views on the possibilities of genetic engineering?  What do you think the American public would say if they heard Abbott, Upjohn, Pfizer, Hoffmann-La Roche, Lilly, say to the American public we companies have a right to patent new forms of life?  How can a company claim the right to patent a new form of life?  What does that mean ten years from now if they can patent a new microorganism today?  What does that mean twenty years from now?  Go back and remember the quotes of Bonner, Watson, and Lederberg.

Let me finish up with one interesting observation:  You ain't seen nothin' yet.  The press here, the critics, think that this is a question of the public interest groups versus the scientists.  Wait until the Protestants, the Jews and the Catholics, the Methodists, the Presbyterians and the Baptists all over America start to realize the long-range implications of what you gentlemen are doing here to-night.  I think that it is time to open this up to public debate.

Now, I am not the only person here.  There is someone here at the microphone now.  How many other people have something they would like to add to the agenda?  Any hands?  Why don't you just come on up to the microphone and let us talk.  Does anyone have anything to add to the agenda?  Let's hear it.  That is all I have to say.  Let's open this conference up or close it down.

HAMBURG:  Let me call to your attention that there are multiple path-ways into this program.  We have tried as hard as we could to put flexibility into it.  It is not just a matter of a series of speakers, however distinguished, scientists and nonscientists.  It is a mat-ter of participation of as wide an audience as we can possibly manage.

First, there will be multiple discussion periods, and we are go-ing to call on just as many people as we can.  Naturally, with the scope here we have to ask you to keep your comments brief, but we think many, many people will be involved in that way.

Second, we have a Panel for Inquiry, seated here in the front row, whose task it is to put searching questions, to probe for the weak spots, to point up contradictions, to bring out the doubts and concerns that have not been adequately brought out by the speakers.

Third--and this is really very important--I would urge you to take advantage of the workshops.  There are quite a number listed, and many of you have already indicated your preference for existing workshops.  But there is the perfectly good possibility of creating additional workshops, and we will do so if there is sufficient in-terest.  So if there is a topic that you care deeply about--and I am very sympathetic with the deep concerns that we all feel in one way or another about this matter--leave us a note about a workshop topic you would like.  We will try to put them all together by noon tomorrow and see if there is sufficient interest to create additional workshops.  So that is a very feasible way to get in any topic which you feel has so far been neglected.  We hope the workshops will

22

make possible active participation by a lot of people who otherwise
might not take part.  These workshops, each and every one, will be
reported to a plenary session on the last day.

Then there is the matter of the publication.  For the much, much
wider audience and for the longer term, the publication is really
more significant than the meeting.  We are going not only to have a
summary of the discussions here with all points of view represented,
but also we would be glad to consider carefully any special submis-
sions that anyone would like to make.

I would also urge you to make informal contact between sessions,
at lunch or in the evening or whatever, with speakers or other people
in the audience you know are active in this field.

Now, having listed all those pathways and all that flexibility,
I think you can see that this is much more than the ordinary program
and that we are trying to broaden out the range of considerations.
It may lead to quite a cacophony of voices, and I think that is quite
inevitable with so many different viewpoints.  All I can say about
that is what Mark Twain said about Wagner's music:  it is not as bad
as it sounds.

In other words, we will try in some way to put together a reasonably coherent, integrated volume that nevertheless, clearly, explicitly expresses every point of view, including those that are
mutually exclusive.  It may be that a certain amount of reconcilia-
tion will be possible in this meeting, but in any event, I have no
doubt whatsoever that the diverse views will be very well expressed.

We did have a chance, Dr. Koshland, Dr. Rich, and I and some of
the others on the Program Committee, to consider this question of
whether the agenda could be changed at the last minute, and we came
to the conclusion that we cannot do that.  People have come from all
over the United States and all over the world for this meeting as
stated, and it will go forward as stated, with all of the flexibility
that I have just outlined, which gives multiple opportunities for anyone who really is serious about it to take part and get their views
on the record.  I am sorry it can't be perfect, but we do think it
goes beyond most meetings of this kind that we are familiar with.

We believe that for all practical purposes, every extant viewpoint will appear in the discussions and in the publication.  Some
viewpoints that haven't even been created up to now will probably
be generated in the course of the meeting.  At any rate, we are
making a very deliberate and systematic effort to do that.  Some of
you feel strongly about one aspect of the problem; there are other
people who feel equally strongly about some other aspect of the
problem, and who see the same issue you do in a very different way.
We are trying as best we can to accommodate this multiplicity of
views.  I think we simply have to try to be tolerant with each other
about these genuinely difficult issues.

AUDIENCE:  I think you should say something about why the structure was
set the way it is before we assume that it should be gone on with.

Jeremy Rifkin referred to the religious community.  It is clear
that the only communities in this society that look at time in
thousand-year chunks are the religious communities.  The whole enterprise of modern science is--

HAMBURG:  I understand your point of view.  Let me say a word about
that--

AUDIENCE:  --this country is only 200 years old--

HAMBURG:  Let me respond.  I am trying to respond to that.

AUDIENCE:  Can you explain why it was not possible to seriously involve
the religious communities, not on the basis that tonight or tomorrow
morning you could sign up to have a little talk to yourself in some
little room, but that they would be together with the panel.  Can
you explain why that was not possible?

HAMBURG:  I will say a word about the way the Program Committee worked,
and then we are going to go ahead with the evening's program.
   The Program Committee itself was very diversely composed of a
wide range of people with different views.  Nobody on the Program
Committee felt entirely satisfied with what emerged.  Certainly I
don't.  I have a number of dissatisfactions.  But it was a compromise among people who not only were diverse in themselves, but who
consulted others very widely, including the religious community.
   There were no strings attached to any of the funds given.  In the
tradition of this series, people who give to it, foundations, corporations, or whatever, know that there are no strings attached.  That
applies to all potential sources.  We did try to contact people who
have every kind of view we could think of.
   Now, if you are dissatisfied that we didn't get your view as
strongly represented as you might have wished, that is very un-
derstandable, but it is not correct to say that we ignored the religious community or any substantial segment of American society.
   We will now go ahead with the Forum as planned.

## THE INVOLVEMENT OF SCIENTISTS

Maxine Singer

*Head, Nucleic Acid Enzymology Section, Laboratory of Biochemistry, National Cancer Institute*

It is almost four years since the morning in New Hampshire when, as co-chairman of the annual Gordon Conference on Nucleic Acids, I said to my colleagues:  "We all share the excitement and enthusiasm of yesterday morning's speaker, who pointed out that the scientific developments reported then would permit interesting experiments involving the linking together of a variety of DNA molecules.  The cause of the excitement is twofold.  First, there is our fascination with an evolving understanding of these amazing molecules and their biological action.  Second, there is the idea that such manipulations may lead to useful tools for allevi-ation of human health problems.  Nevertheless, we are all aware that such experiments raise moral and ethical issues because of the potential hazards such molecules may engender....  Because we are doing these ex-periments, and because we recognize the potential difficulties, we have a responsibility to concern ourselves with the safety of our co-workers and laboratory personnel, as well as with the safety of the public.  We are asked this morning to consider this responsibility."

As a result of the discussion and vote later that morning, a letter was sent to the presidents of the National Academy of Sciences and the Institute of Medicine from the participants in that meeting, some of whom were the pioneers in recombinant DNA research. That letter and its publication in *Science* Magazine initiated a series of events of which this Forum is the latest.

Perhaps most significant was the publication, in 1974, of the report of the Ad Hoc Committee on Recombinant Nucleic Acids, a group that included molecular biologists who were actively pursuing recombinant DNA work. Their report established certain precedents that have been central to all of the activities on recombinant DNA since that time. Thus, the report defined the possible hazards to include the effects on human and nonhuman living things. The report called for an international discussion since the potential hazards could not be limited by national boundaries. The 1974 report recognized that the deliberations could not remain ad hoc, but needed to be assumed by proper governmental bodies which represented the interests of society at large. The ad hoc committee recognized that for reasons of safety certain experiments ought not be done, at least for the time being, and called for their colleagues around the world to join them in a deferral of those experiments. And finally, the committee established the precedent that the discussion must be open and publicized.

Much of what has happened since 1974 has been in response to the request made by the ad hoc group. The Asilomar Conference in February 1975 was the first attempt by an international group with varied expertise to look at many types of recombinant DNA experiments, and try to rank them as to potential danger. The Asilomar recommendations again advised that certain experiments ought not be carried out and, for other experiments, attempted to define levels of containment appropriate to the estimated risks.

In this country the NIH alone among governmental organizations early assumed responsibility for serious and sustained consideration of the problem. The NIH effort resulted in the publication, in June of 1976, of "Guidelines for Research on Recombinant DNA." The NIH guidelines are based on analyses that are similar to, but more detailed than, the Asilomar review, and have explicit containment requirements for most technically feasible experiments.

Publication of the guidelines was not, as some have implied, a "go" signal for all recombinant DNA research. Contrary to public belief, the voluntary deferral that started in the summer of 1974--it has been referred to as a moratorium--did not call for a ban on all recombinant DNA research. Only two types of experiments were deferred: first, the construction of drug-resistant or toxigenic microorganisms that do not occur naturally; and second, the introduction of all or part of the genomes of viruses known to cause cancer in animals into bacterial cells. At the present time there are no viruses known to cause cancer in humans.

But there are many other types of recombinant DNA experiments that are feasible and important, and their potential for hazard is not clear-cut: they are not covered by the deferral. In the Asilomar recommendations and in the NIH guidelines, the experiments deferred in 1974 either remain proscribed or can be performed only under extremely stringent containment measures. The guidelines forbid additional experiments, including many that have provoked great fear of the possible hazards of recombinant DNA research in the mind of the public.

From July 1974 until Asilomar in February 1975 and from then until the publication of the guidelines in June 1976, there was, as far as can be learned, complete compliance with the then-governing prohibitions and containment recommendations. Experiments that were not prohibited were carried out during the entire period. There is, thus far, no indication that hazardous organisms have resulted from any of the experiments. Indeed, with the exception of certain experiments involving antibiotic resistance and toxins, we still do not know that hazardous organisms can in fact be produced from recombinant DNA experiments. We cannot accurately describe the probability of or the precise nature of the conjectured hazards. Statements implying that uncontrollable epidemic or environmental disaster is a certainty are as misleading and useless as statements implying that no possible hazard can come from these experiments. Insufficient knowledge is the reason why the public is faced with a range of different opinions from within the scientific community. Insufficient knowledge is also the reason why the recommendations in the NIH guidelines were necessarily based on judgment and consensus.

The adequacy of the containment requirements mandated by the NIH guidelines for permissible experiments is a useful focus for discussion. In this way the very different issues raised by different experiments can be considered. Misleading and sweeping statements referring to all recombinant DNA experiments can be avoided. Most scientists and laymen who have studied the situation agree that certain recombinant DNA experiments that mimic naturally occurring processes are without unique potential for harm. Most people agree that certain other experiments ought not be done at all at the present time. The facile description of people as either "proponents" or "opponents" belies broad areas of agreement, as well as the complexity of the issues. Similarly, the facile description of bacteria containing recombined DNA from a foreign source as "new living things" is misleading. A bacterial cell normally contains thousands of genes, each of which contributes to the nature of the cell in interdependent ways. The introduction of one or a few new foreign genes to this complex system may be able to alter certain properties of the cell, but the bacteria basically remains its old self.

Questions do remain about certain specific recommendations in the guidelines, and the need for additional or different provisions is a subject of debate. For example, the current controversy over whether or not recombinant DNA experiments may cause long-term evolutionary consequences is properly part of the debate on the provisions of the guidelines since the risks are imagined to result from a particular type of recombinant DNA experiment.

Debate aside, there has been substantial endorsement of the NIH guidelines both within the scientific community and by responsible representative public bodies including the Cambridge City Council, the University of Michigan Board of Regents, and the Senate Subcommittee on Health. All work supported by federal government funds is now covered by the guidelines; they are viewed as mandatory by grantees and grantors. The threat of removal of research support is a powerful sanction, not a trivial one. Institutional biohazards committees are functioning at grantee institutions and at the NIH. Reports from the committees indicate a diligent and serious commitment to the provisions of the guidelines. Most dramatic evidence of this compliance comes from the willing destruction of materials constructed in accordance with the Asilomar recommendations but prohibited by the NIH guidelines, and from the straightforward discussions of risk and containment now appearing in published scientific papers.

There remains an urgent need to extend the provisions of the guidelines in an enforceable manner to work carried out with nonfederal funds. The NIH does not have such enforcement authority and, as a principal research sponsor, is not an appropriate agency for such a task. Intensive federal efforts to find suitable enforcement mechanisms are nearing completion, and we may expect to hear about these efforts this week. Discussion is also proceeding actively in several state and local governments.

The current situation in the United States is but one aspect of worldwide attention to this problem. The scientific community, working through its extensive international collegial network, sought and obtained official attention to the problem of recombinant DNA in many countries. Two countries, Canada and the United Kingdom, have independently developed guidelines; although they differ in detail from one another and from the American guidelines, all three agree in general approach, and to a large extent in the assessments of relative risk. Other countries will make use of one or another of these sets of guidelines, organizing the implementation of them in ways appropriate to national conditions. Several international organizations--on the official governmental level, the World Health Organization; on the scientific level, the European Molecular Biology Organization and the International Council of Scientific Unions --have active programs designed to foster both science and safety by collection and dispersal of information and by training of investigators.

Scientific progress with recombinant DNA techniques has been slow. Meeting the requirements of the guidelines--from prior approval and certification before initiating experiments, to the demanding containment requirements--has slowed the pace of work. Certain permissible experiments are not presently feasible because of the lack of required physical facilities or the lack of appropriate certified hosts and vectors. The committee advising the Director of NIH on certification of biologically contained hosts and vectors has been rigorous in its evaluations.

This slowdown is useful. It allows time for prudent evaluation of the accumulating experimental results, and the implications of those results relevant to potential hazards. The slowdown is also frustrating not only because it has delayed acquisition of information, but because research is a creative as well as a technical endeavor. In successful, innovative work, the impetus of enthusiasm, of acting quickly upon an exciting idea, is undeniable.

As I mentioned before, certain recombinant DNA research has continued during the last few years. Those experiments have confirmed the initial enthusiasm for the value of the method. It is now known that the DNA of higher organisms, from yeast to mammals, can be faithfully reproduced in bacterial cells, and DNA of bacterial origin is readily reproduced in animal cells, growing as single cells in tissue culture. Thus, the promise of the method for the preparation of useful and otherwise unobtainable quantities of specific DNA fragments or genes is an established fact.

Transcription of the information encoded in DNA into RNA is the first chemical step in genetic expression. We now know that the DNA of complex organisms can also be transcribed into RNA inside bacteria; and similarly, the DNA of simple organisms can be transcribed into RNA in cells derived from complex organisms. Detailed study of these systems promises the elucidation of important questions concerning the control of genetic expression.

The final step in gene expression is the translation of the information in the RNA that results in the formation of a protein. Ultimately it is the set of proteins unique to each organism that define the recognizable properties of each species and each individual. Proteins encoded by the DNA of yeast, a primitive form of higher organism, are synthesized and are active in bacterial cells. These results indicate that some initially speculative practical applications of recombinant DNA techniques will be realizable. Taken all together these results confirm the unity of nature both in structure and function.

It was not easy for the scientific community to raise the issues implicit in recombinant DNA research. The actions involved significant divergence from historical practice and belief. The actors were unaccustomed to consensual undertakings, and the wisest course was not clear. Doubts still persist about the wisdom of each step that was taken. Those colleagues who warned that uncontainable and irrational public responses might follow were correct. But their counsel was set aside because other considerations were overriding; and it is worth making these other considerations explicit.

Scientists today recognize their responsibility to the public that supports scientific work in the expectation that the results will have a significant positive impact on society. To describe the scientific community of the late twentieth century otherwise is to ignore or to misunderstand the evidence. Dispute over the best way to exercise that responsibility must not be confused with the negation of it. The scientific community has accepted the counsel of ethicists, philosophers, and representatives of the public who long troubled to point out this responsibility. Origins for the actions regarding recombinant DNA are also found in the worldwide movement to protect the biosphere from the ravages of technological development. And again, while we need continuing discussion of the proper balance between efforts to ensure environmental protection and opportunities for solutions to existing and forthcoming problems, we all agree about the importance of environmental considerations.

Scientists also accept the need to restrict certain laboratory practices in order to protect the safety and health of laboratory workers and the public. Further, we recognize the need to consider possible hazards before large-scale activity is undertaken, and before untoward events occur. But we differentiate between restrictions on hazardous or potentially hazardous activities, and restrictions on intellectual freedom. While a democratic society rests on the virtually absolute freedom of individuals to ask any question whatever, it is clearly unacceptable knowingly to cause harm to others in the process of trying to obtain an answer. Thus the recombinant DNA problem was originally posed and has been dealt with as a problem in the safety of living things.

Some have argued that this definition of the problem was too narrow. It is said that scientists and the public should consider the moral and ethical implications of future applications of the knowledge to be acquired from this research. And so they should, but in broader contexts and with even wider participation then was engaged in dealing with the technical matters of safety and laboratory practice.

Further, it has been argued that scientists should not only consider, but should in fact assume responsibility for the eventual application of any knowledge they may acquire in the course of research. That statement raises complex and difficult issues and varied responses. It can be a subject for reasonable debate only if the distinction between acquisition of knowledge and application of knowledge is not obscured. Thus any exercise of such responsibility can logically come only after the acquisition of the knowledge. To call for such an exercise prior to the research itself is a sham, because the outcome of the research is, by definition, not knowable in advance. This is not to say that freedom of inquiry is unlimited; but limitations on the acquisition of knowledge must be with good cause, as when harm may result from the process of acquisition. History reminds us that constant vigilence is required if we are to avoid the perilous consequences of attempts by society or by individuals to determine what is permissible to know and what is illicit to learn. The consequences of attempts to restrain the search for knowledge have been even more fearsome than the science fiction scenarios constructed by genetic fearmongers.

Besides, such attempts are certain to fail. They will fail, first, because we are not smart enough to foresee what we will or will not learn from a given line of research. They will fail, second, because we are not smart enough to foresee all the future applications of the knowledge. They will fail, finally, because the indomitable forces of nature oppose such attempts. The acquisition of knowledge by the human brain is part of protean nature. Biologists and poets alike know this. Emily Dickinson wrote, in 1862:

> The Brain is just the weight of God--
> For--Heft them--Pound for Pound
> And they will differ--if they do--
> As Syllable from Sound.

Most scientists today also recognize the need to participate, together with the public, in decisions about research areas ripe for encouragement, or areas where knowledge is desired, or areas in which safeguards may be needed. The worthy report of the Cambridge Experimentation Review Board must surely quiet doubts about the ability of the lay public to deal intelligently and forthrightly with complex technical issues. Future public reviews of such matters will be judged against the standard set by that Cambridge Review Board. But cooperative deliberations between scientists and public bodies is difficult because scientists have not educated others adequately in the past. It should not then be surprising if deep fears and ambiguities arise in the minds and hearts of those who suddenly learn the depths of modern insights into the nature of living things.

On the other hand, those responsible for making public policy should recognize that levels of anxiety are often unrelated to levels of risk. A continuing search for effective means to inform and educate the public about science is essential. The history of the debate over recombinant DNA suggests that current means give erratic results. Press coverage of the Asilomar Conference in February of 1975 was excellent. As public discussion broadens, however, we encounter serious problems in the presentation of the issues and the science to the public. Communication between scientists and the public is often impeded by writers or TV producers who unfortunately take it upon themselves to determine what the public needs to know, or what the public can understand. The public is the loser as they are inadequately or incorrectly informed. And serious ethical considerations confound the scientist if his efforts to cooperate with the media are used to misinform or needlessly frighten.

For the future, scientists need to continue, together with federal and local governments, to evolve policies that offer protection from potential hazards, and preserve opportunities for discovery and development of safe and desirable applications. Scientists must share their insights into the nature of living things with increasing numbers of people so that debate can be predicated on understanding rather than on fear. In order to counteract the growing pessimism about the nature of knowledge, the proper separation of science from technology must be made, and in the continuing dialogue the distinct values and problems inherent in each must be carefully articulated.

Finally, if scientists commit themselves to their unique opportunities to serve as an early warning system, society can proceed with prudence and caution as scientific knowledge grows.

# THE INVOLVEMENT OF THE PUBLIC

Daniel Callahan

*Director, Institute of Society, Ethics and the Life Sciences*

The relationship between science and the public has often been likened to
a romance. More recently it has been said that the romance is now coming to
an end. But if the former view was correct, I doubt that the latter is
quite accurate. On the contrary, the romance continues, but the difference is that the partners are now beginning to live together. That always
adds a few well-known complications: more intensity, but also more fights,
the potentiality of greater devotion but also of greater disillusionment.
Romances conducted at a distance always go more smoothly. Problems begin
only when bodies come together--and of course that is when real possibilities begin as well.

We are now gathered here in a public marriage counselor's office.
All we lack, unfortunately, is a wise marriage counselor. Perhaps he or
she will turn up. In the meantime, we must display both our mutual loves
and hates in public, even before the press.

The case in question has many familiar elements about it. There are
charges that one of the partners only married the other for money. And
there are charges, and countercharges, that the real motives and interests
of the partners are other than what they claim them to be. There are also
mutual accusations from both sides about the unwillingness of the other
side to talk openly and frankly. "Why won't you talk with me," they
each say. To which the other responds, "but I try and you just don't
listen." "But when did you try?" goes the response. "During our trip
to Asilomar," is the answer. "Are you kidding," the other replies. "Well,
I certainly tried in Cambridge, and in New York, and in Sacramento," is the
retort. "You call that talk; we just yelled at each other."

I think we have already heard enough of that too-familiar marital
dialogue. Our task is to see if we can sort out the problems and to tran-
scend the petty and destructive way two parties in conflict can destroy
each other. Let us begin at the beginning and see what we have here.
What has been the relationship between science and the public on the issue
of recombinant DNA?

Though there were earlier informal events, I will date the beginning
as the Gordon Conference in the summer of 1973. The outcome of that conference was a letter to the National Academy of Sciences and the Institute
of Medicine requesting the establishment of a committee to study various
problems of recombinant DNA research and to recommend specific actions or
guidelines in the light of potential hazards. That was and remains a
striking act of moral initiative and courage. A committee was formed, and
it recommended that a moratorium on certain forms of recombinant DNA research be voluntarily established. It also recommended that the National
Institutes of Health (NIH) set up an advisory committee to, among other
things, evaluate potential hazards in the research, to devise safety

procedures, to develop guidelines for researchers working with potentially hazardous DNA molecules, and to call an international conference. In October 1974 the NIH established the Recombinant DNA Molecule Program Advisory Committee. In February 1975, the international conference earlier called for was held at the Asilomar Conference Center in Pacific Grove, California. In essence, the conference participants concluded that the voluntary moratorium should be lifted and that in the future research should be conducted under a set of guidelines. Immediately thereafter, the NIH Advisory Committee began work on refining the guidelines suggested at Asilomar.

Now where, up to this point in the history of the matter, was the public? The only straight answer is that the public was not much in evidence. Save for the presence of four lawyers at Asilomar and the presence of a good number of reporters, the public was at that point little involved. What are we to make of that fact? Nothing very portentous, I want to suggest, and certainly nothing deceptive. The very first thrust of the recombinant DNA debate was within the scientific community. The group now called the Berg *et al.* group wanted to raise a moral issue in the scientific community: some of the proposed recombinant DNA research *could* be dangerous. They succeeded in putting that moral issue on the table, and not without opposition. They recognized the significance for the public of what was, at that stage, a struggle among scientists trying to determine whether there were or were not real dangers in the research. They signaled that fact quite openly and clearly by inviting the lawyers and the press to Asilomar.

It is very easy to second-guess the whole procedure up to that point. Surely, one might say, a handful of lawyers hardly constitutes a full involvement of the public. And surely too, press coverage of what was otherwise an essentially closed event is something less than public participation in scientific decision making. True enough, but that is all retrospective wisdom. In the first place, the whole problem was new. There exist no real historical precedents, including the development of the atom bomb, for a problem of this kind: What ought one to do in a situation of spectacular hypothetical possibilities over against equally spectacular hypothetical dangers? In the second place, there did not then nor does there today exist any ready-made forums for public discussions of matters of this kind. The scientists had to create their own public forum. Had they not done so, we would probably not be here today, and there would probably never have been debates in Cambridge, Ann Arbor, New York, and elsewhere.

But let me return to the history. During 1975 the NIH Advisory Committee worked hard to develop guidelines, and in December of that year a draft set was sent to the Director of NIH. An obvious question arises at this point. Were there any representatives of the public on the NIH Advisory Committee? No, there were not at that time. Though a lay person was added later, that was a real oversight. Moreover, it was essentially a technical and not a policy committee. The next important event was the calling by Donald Frederickson, Director of NIH, of a meeting of his Director's Advisory Committee. To that meeting, held on

February 9-10, 1976, were also invited representatives of various public interest groups, representatives of various factions within the scientific community, and assorted other people.

The purpose of the meeting was a public examination of the draft guidelines. They were fully explained and then debated by the participants. On the whole, it was a subdued debate. The majority of the participants, including myself, felt they were in general satisfactory. The strongest opposition came, not from the public representatives (with the exception of Susan Wright from the University of Michigan), but from other scientists, notably Robert Sinsheimer and a group of young scientists from Boston. Peter Hutt, former General Counsel of the Food and Drug Administration, raised what for me was the crucial question: Is the burden of proof to go ahead with recombinant DNA research on those who support the research, or on those who oppose it (or at least want much tighter guidelines)?

The implicit answer, I believe--for the question was not really taken up--was that the burden of proof lay with the opponents. That is hardly surprising either in this or in most other instances of technological innovation. We live in a country which still has a great faith in scientific progress, and an equally great dependence on such innovations for our economic, medical, and social well-being. In that general cultural situation it seems pointless to blame individual scientists for wanting always to move forward. If there is a fault, it lies deeper than that in our society.

However low-keyed the February 9-12, 1976, meeting at NIH, skillfully and fairly run by Dr. Frederickson, the ingredients of a coming storm were present, not fully to be exploited until later. There was, for one thing, a disturbing realization on the part of some that the debate and discussion on guidelines had moved very fast. Why was it, for instance, that the first fully public discussion of recombinant DNA was already focussed on draft guidelines? A decision even to prepare and discuss guidelines seemed to assume that the basic ethical, political, and social questions about recombinant DNA research had all been raised and fully discussed. But they had not been. For another thing, while there were certainly public representatives at the NIH meeting, and many reporters as well, it was hardly likely in such a heavy and genteel setting that the full voice of the public could have been heard, or that, if heard, it could have significantly slowed down the rapidly moving machinery of putting guidelines in place to govern NIH grants. It should hardly be astounding that some felt vested interests were pulling the strings and that, once again, the scientific establishment was slipping one over on the public.

I reject that cynical interpretation. The immediate problem of NIH was to set some ground rules on a form of scientific research that was moving very rapidly and that, one way or the other, had to be controlled and monitored. The guidelines did and do represent a compromise solution. But, some have complained, in such a potentially fateful situation, compromise is not good enough. They might be right--future generations will make that judgment--but the reality of matters now is that neither the public nor the scientific community share any consensus whatever on recombinant DNA research. And what do we do in our society when there is no consensus on an issue? We normally compromise, for that is the one way we can continue to live together and at the same time keep the debate going. If NIH had not acted as rapidly as it did, one can be almost certain that the research would be going on at a much faster pace than it is today, and that the public would know even less than it does now. If NIH had decided to declare a long moratorium on all recombinant DNA research, my own guess is that it would have gained the support neither of the public nor of most segments of the scientific community. It could not have made such a moratorium stick.

Nonetheless, the fact remains that up to and including the February 1976 NIH meeting, there was very little public participation in the recombinant DNA debate. That was soon to change. As the headline of a February 11, 1977, article by Nicholas Wade in *Science* put it, "Gene-Splicing: At Grass-Roots Level a Hundred Flowers Bloom." New York State held public hearings in the fall of 1976 on recombinant DNA research, and a bill has been introduced into the New York legislature to regulate the research. The same thing has happened in California, and is being considered in New Jersey. In Madison, Wisconsin; Bloomington, Indiana; and San Diego, California, discussions and hearings on the issue have taken place. A major debate took place during 1976 at the University of Michigan, resulting finally in a six to one vote on the part of the Regents of the university to allow the research there to proceed. The most flamboyant public squabble of all took place in Cambridge, Massachusetts, where the city council, led by Mayor Vellucci, imposed for some months a moratorium on the building of a recombinant DNA laboratory at Harvard. The moratorium was only lifted after a special citizen's commission approved the construction.

Meanwhile, also during 1976, a number of environmental groups became actively involved with the issue. The Environmental Defense Fund, the Natural Resources Defense Council, Friends of the Earth, and the Sierra Club have all, in one way or another, taken an exceedingly dim view of the way the public has been involved in the debate and have, with minor variations, called in effect for a new public debate. Senator Dale Bumpers has introduced a bill in Congress to regulate recombinant DNA research, and Senator Kennedy is expected to hold hearings before the Senate Health Subcommittee.

So, in one way or other, the public is now involved. But there remain of course a few questions, and I want to focus on three that seem to me central. The first is: Why did it take the public so long to get involved and what were the circumstances that explain the development of that involvement? The second question is: Now that the public is involved, what options are open to it and how might they be evaluated? The third question is: What ethical and social criteria should the public use in judging and deciding upon the future of recombinant DNA research? As it will turn out, these questions are not unrelated to each other.

1. *Why did it take the public so long to get involved in the issue?* It does not seem to me adequate to say it was because scientists kept the

public out.  They did not.  Even if it is true that wide press coverage
is not public participation, it is surely a necessary condition for that
participation.  The Asilomar Conference was well covered in the press,
and there were frequent follow-up stories throughout 1975 and, as the debate heated up, even more in 1976.  If the public had wanted to jump in
right after Asilomar, it could have done so.  But the public did not leap
at the opportunity.  Why was that?  There is one obvious explanation.
The lag time between a complex scientific issue being raised in public
and public interest in that issue is almost always fairly long.  Yet I
thing a more subtle and supplementary explanation is necessary in this
case.  My own theory is that the public did not take a real interest in
the issue until some senior and notable scientists entered the fray, and
entered it in a very outspoken way on the side of the doubters.  I am
referring to the advent of Erwin Chargaff, George Wald, Ruth Hubbard, and
Liebe Cavalieri on the scene, which occurred during 1976.  To be sure,
Robert Sinsheimer had gone on record earlier with his own doubts; and
Science for the People in Boston, primarily composed of younger scientists
and those interested in science, risked and got considerable wrath for
publicizing their own hesitations.  They deserve very special praise.
But none of them commanded the kind of public attention that Chargaff,
Wald, Hubbard, and Cavalieri did, if only because of their scientific
eminence and seniority.

Why was that important?  For one obvious reason.  It is very hard,
if not impossible, for the public to get interested in scientific decision
making unless potential social and ethical issues are called to their
attention by scientists.  It is even more difficult to mount a full public
debate unless the public has some scientists to lead them into battle.
So it was on the early environmental struggles and so it was on the debates over nuclear energy.  In this case, as Mayor Vellucci was quoted
in *New Times* (in his characteristically understated way), "If I'm gonna
take a stand against this goddamn thing, I need some people on *my* side.
And since they said they would come [Hubbard and Wald], I was fortified,
I was ready for a meeting, and that is the reason why we then flung the
challenge at Harvard and MIT to send *their* scientists over here because
I knew *I* had scientists on *my* side!"  Whether their worries and arguments
are right or wrong, then, the advent of Chargaff and the others provided
a scientific rallying point for those members of the public who wanted
to know if their own hesitations had any scientific basis.  Just why those
scientists were not heard from at the time of the Asilomar Conference and
in 1975, I do not know.

There is still another point to be made about the public involvement.
As I mentioned earlier, there still does not exist any natural forum for
debate of issues of this kind.  If the public had displayed an early
interest, where could they have gone?  As the debate within the scientific
community developed, there were no public organizations in place ready
to grab the issue and run with it.  Though various environmental groups
are now involved, groups which are skilled in publicizing issues and
developing grass-roots interest, they are not necessarily the best-
equipped organizations to deal in the long run with the complex ethical,
social, and legal questions at stake.

   2.  *Now that the public is involved, what options are open to it and
how might they be evaluated?*  A number of options have been proposed, and
they can be classified into two groups.  The first and most moderate
would be to turn the present NIH guidelines into state or federal law,
perhaps modified, perhaps not.  The advantage of that approach is that
it would make up for the most obvious and necessary deficiency in the
NIH regulations--that they apply only to federal grantees.  They do not
apply to those doing recombinant DNA research under private grants and,
most importantly, they do not apply to the research of private industry.
Moreover, they are only guidelines, lacking the force of statutory law.
In my own view, there probably should be such laws, and preferably
federal rather than state laws.  More to the point, a public debate over
whether there should be laws governing recombinant DNA research would
have the healthy effect of allowing a significant public participation.

   The other options would go considerably farther.  They would, in
effect, mean scrapping the present guidelines and beginning the whole
debate over again.  Two groups, the Environmental Defense Fund and the
Natural Resources Defense Council, have petitioned the Department of
Health, Education, and Welfare for hearings to determine if any recombinant DNA research should be allowed to go forward, and under what conditions.  Friends of the Earth goes still farther, demanding a moratorium
on hazardous experiments pending the outcome of congressional hearings.
Susan Wright, of the University of Michigan group, has called for a complete moratorium on all recombinant DNA research "until policy options
have been carefully considered and chosen through democratic procedures
developed for the purpose."  Robert Sinsheimer, George Wald, and Erwin
Chargaff would allow the research to be done in only one national laboratory, and then under very strict controls.  Clifford Grobstein has offered a middle-ground proposal.  He asks that a joint commission,
appointed by the President and by Congress, be established.  Its task
would be not only to take up the ethical, social, and legal issues
earlier skirted, but also to analyze all aspects of the problem.  A
full assessment by the commission would be due not later than two years
from its initiation and not more than three years from the date of the
NIH guidelines.

   Two brief points.  I do not think a full and total moratorium is
possible, even if it might be desirable.  Such a flat ban on research
would require a consensus that, as suggested earlier, simply does not
exist in our society.  It would also beg the question of whether the
research is potentially hazardous.  Nor, for that matter, does there
exist any consensus on the need for caution and slow movement that would
have to be the underlying value premise of such a moratorium.  I also
have doubts about concentrating all research in a single facility.  I
am not certain that is the best way to get the best science, and at
least we want that, and I am even more skeptical whether such a necessarily well-guarded and quarantined facility would be a good starting
point for full public disclosure of all the facts as the research moves
forward.  Full disclosure is considerably more likely where many facilities are doing the research and where many other scientists, involved

in the research or not, are able to watch the process.  It stretches my imagination a bit to suppose that the sharpest critics of the research would be offered positions in a one-and-only national recombinant DNA facility.

Will it be possible to develop the democratic mechanisms for full public debate now being called for?  I believe so:  first, by means of the debate necessary to turn the NIH guidelines into federal law, making them applicable to all who do the research; and second, by the establishment also of a federal commission, as Professor Grobstein has suggested, to fully examine the issues.  Or, alternatively, if not a new commission put together for that purpose, to make use of the present National Commission for the Protection of Human Subjects of Biomedical and Behavioral Research for that purpose, which has already established a solid reputation for seriousness, diligence, and fairness.  In the meantime, speaking as at least one member of the public, I am well prepared to live with the present NIH guidelines.

3.  *What ethical and social criteria should the public use in judging and deciding upon the future of recombinant DNA research?*  This is to me the most fundamental question.  Though I think that the course chosen by the original group that signed the letter to the National Academy of Sciences and the Institute of Medicine and by the NIH in developing its guidelines, given the novelty of the issue and the need for quick action, was eminently defensible and worthy of praise, the critics are still correct in a very *general* way.  There has yet to be a good national public discussion.  Worse still, from what I can see, the discussion is not getting any better.  It is, in fact, getting boringly repetitive in substance and tediously hysterical in tone.  It is simply not enough to affirm the high principle that the public should have a role in the decision making, that there should be public forums and public debates.  What ought to be the *content* of that discussion?  By what *criteria* ought the public to judge the competing scientific and ethical cases that have been made by now?  For the public should not only be heard.  The public ought also to think.  But what is it supposed to think *about*?

I can only make some suggestions here on that question.  I have not thought about it enough myself to do more than that.  First, I believe the public needs to think most carefully about the whole idea of scientific progress.  As a general policy, does it favor nerve, boldness, and risk-taking, or does it favor caution and slowness?  Which is the wisest future direction of our public policy in this respect on basic research?  And what counts as wise?  Second, I would like to know whether the public, after considerable thought, thinks that we have a *moral obligation* to pursue lines of research that may benefit present and future generations?  I say "moral obligation" because it is sometimes implied by advocates of recombinant DNA research that science would be guilty of a sin of omission if it did not continue and promote research so promising in theoretical and practical benefits.  I myself would prefer to say that the research is desirable and valuable, but by no means is it morally obligatory.  It is just one choice among many we can make in allocating our scientific resources.  But I would like to know what the public--after due consideration--thinks about all that. Third, what does the public think about risks and benefits? How, in some rational way, *ought* the public to think about that problem?

One obvious implication of this line of thinking is that the public has as much obligation to act responsibly as does the scientific community. All of the appropriate fuss being made about the need for socially responsible scientists could well be matched with some concern about a socially responsible public. The future of the recombinant DNA debate will turn on the quality of the dialogue between the scientific community and the public. Neither side can conduct the debate on its own. Each needs what the other has to give. That means the public must be kept informed in the future, must have a central role in present policy formation, and must develop standards by which to judge the issues. It also means that scientists must bring their knowledge and, just as importantly, their lack of knowledge in this case out into the open, not once but again and again.

We have, the public and the scientific community, now begun to talk. This marriage can be saved.

## DISCUSSION

JONATHAN KING, Department of Biology, Massachusetts Institute of Technology: I am here representing Science for the People. I would say that both the presentations we just heard constituted not what it says in the program, but essentially a kind of whitewash. Dr. Singer, whose initial role in this whole issue was a very productive one, has essentially come up here and described to us what I would describe as a technocratic coup by a small sector of the scientific community acting under the guise or in the name of scientific responsibility. This has been presented as the best of all possible worlds with those who criticize on the one side or those who criticize on the other side as a lunatic fringe who will be swept away under rational progress. It has been suggested that what has been done is what should have been done, and that is the way we should proceed.

To repeat Mr. Callahan's query: Why weren't the critics at the Asilomar Conference? The Walds, and the Hubbards, and the Cavalieris, and the Beckwiths, and the Kings, and the Signers were not invited to the Asilomar Conference. Moreover, there were no microbial ecologists there, few public health experts, no environmentalists, and no people from the National Institute of Occupational Safety and Health.

Then we have the National Institutes of Health constituting an official committee of fifteen people, again people who are primarily mostly concerned with research, not people who are involved in occupational safety and health, not microbial ecologists, not environmentalists, constituting a group who then kind of officializes the policy. The conflicts of interest are unbelievable. One of the chief proponents of the research, the chief developers of the technology from a West Coast university is made the chairperson of the subcommittee writing the guidelines.

This is then presented as a general discussion. But there is some understanding that it wasn't as broad as it could be, so the NIH has "public hearings." Mr. Callahan mentioned that the public was not well represented. I will tell you why there weren't members of the public there: because they couldn't afford to go. For example, we knew a couple of people at MIT who work in the laboratories and make $7,000 or $8,000 a year. They were quite willing to go down to the NIH and testify about what conditions were actually like in a laboratory, not what professors said they were like, but what they were like. But these people were going to lose a couple of day's pay, and it costs money to go down to Washington. In fact, it was going to cost them about three weeks' salary to go down there for one or two days. And so, for example, a few of us called up the NIH and said could you bring these people down. You have paid for so and so to come in from the West Coast, how about paying for a couple of bottle washers to come down from Cambridge? No, they couldn't do that. So sure, we didn't have much in the way of the public represented.

Now, I would say that the whole procedure of coming up with the guidelines reflects the viewpoint of a sector of the scientific community involved in the research, no more, no less. But to have that become national policy on what is safe or what is not safe in this research, that is not democracy. That is technocracy in the deepest sense: a small group, who because of their situation, that they are involved in the research, find themselves in the position where their word is national policy. There wasn't a conspiracy or anything. I mean, that is just the way it worked out.

Dr. Singer pointed out in her presentation that the way it was was fine: everything is wonderful, we are going to go ahead and lead on to horizons and rainbows. Let me give you an example. Now, as one who is actively or even militantly involved in the debate of dangerous organisms, time and time again I was hit over the head with the fact that we don't know whether genes of even yeast are expressed in *E. coli*. Finally it comes out that they are expressed in *E. coli*, and what are we told? That this opens the way for productive applications. Why weren't we told that this opens the way for taking some of those hazards extremely seriously, because it is true that the proteins of the eukaryotes will be expressed in the bacteria. The point of view expressed was the point of view of those people who were involved in setting up the guidelines. They naturally see the world as what they did was fine and what everybody else is criticizing is craziness.

I think there is a very profound confusion and a terrible misfortune, or misrepresentation made to many members of the scientific community. This is not a question of freedom of inquiry. This is a question of freedom of manufacture, of modifying the environment, of modifying living organisms, not of asking questions about them, but of the route which you take in getting the answer. No one of us is saying don't accumulate knowledge.

I was a graduate student at Cal Tech during the war years, where there were a lot of missile engineers. A number of us were concerned that these people were using their scientific skills to design devices to kill people. And we would raise questions sitting around the dormitory, and they would say you are interfering with our freedom of inquiry. What freedom of inquiry? You are making missiles. They would say we are not making missiles; we are studying the motion of an elongated projectile through a liquid medium, and if we cannot do that we cannot learn about it.

We are being told that if we don't want to have this experiment done, modifying living organisms to have it proved to us that it is not a disaster, we are holding back knowledge. Now I ask you, what is going to happen if by some chance, by some small chance, the Walds, and the Hubbards, and the Chargaffs, and the Cavalieris are right; the experiment is done, and we get the answer—a disaster. Where will we be?

Let us look at the other side. Suppose we have a moratorium and we cut off research on this issue. We cut off the use of this technology. We don't cut off asking questions about mammalian chromosomes and DNA. I am a geneticist myself. I love genes. I love chromosomes. I make my living studying them. But we cut off the use of this technology in which you shuffle, according to the whim of some fashion in the scientific community, the products of billions of years of evolution. Will the pipettes disappear? Will the scientists disappear? Will the labs close down? Will the fount of all knowledge dry up? For crying out loud, no. One small sector of the scientific community will have to do a different experiment for the way to knowledge.

The government and the taxpayers support scientific research to improve the national health. The fact that the scientists don't care that much about it or may not care that much about it is just an unfortunate distortion of the history of the funding. If you go back and read the *Congressional Record*, why did they put that money into national biomedical research? It was to improve the health care system. They couldn't put it directly, they couldn't get national health insurance, they couldn't get a national health service. They said let us slip money into health care through research. Research is a good way to spend money. But money is spent on research in order to improve national health.

It is true that the mammalian chromosome is very important. But a miner in West Virginia who gets black lung doesn't understand the molecular mechanism of why he can't breath, but he knows he got it in the coal mine. And the cotton workers in the textile mills in South Carolina and Georgia who have byssinosis and emphysema because of the cotton fibers they are breathing don't understand the molecular mechanism of it. And maybe we don't understand exactly how all those carcinogens in the environment give us cancer.  But this is a technology that cannot clean up the carcinogens in the environment.  This is a technology that says we give up on the causes of disease.  We give up on preventing disease; we are going to modify the people to make them resistant to the disease.  It is not that individual scientists are doing that; individual scientists are engaged in the pursuit of knowledge and research.  And it is not in the future.  It is in the scientific literature now where the national cancer program has a component that says develop the means to modify individuals to reduce the risk of cancer, and up until a year ago it didn't have a component that says identify carcinogens in the environment.

Biological scientists have had the luxury in the last twenty years to be cut off from the forces of war, to be cut off from the forces of destruction, to be cut off from heavy profit involvement, from technology being twisted to make the last buck out of it rather than in the public good.  But now all of us are in a sphere where that is beginning to happen.  We see it with the drug companies.  We see it with the Cetus Corporation.  We see it with the Genotech Corporation.  We are seeing here the entry into scientific research of the profit motive in a very subtle way.  Don't identify the carcinogens in the environment, but develop this super-fancy technology, very sophisticated, capital-intensive, a million dollar start-up cost to build a safety facility.  All of us scientists, if we want to make sure that the products of the human intellect, that scientific knowledge and scientific skill are used to benefit the people of the country who worked to keep it going, the working people, the people who wash the glassware in the laboratories and make the scientific instruments, and make the pipettes and keep the lights going, then we have to take a much closer look at this whole process than is afforded to us in a forum like this.  We cannot sit by here and hear people say everything is fine.

Dr. Singer didn't mention why it was that at the University of Michigan they had to fly in two of us crazies from Cambridge to testify in opposition, why it was that not a single biologist in the whole University of Michigan was willing to get up and testify.  There was a Science for the People chapter there, and there are a lot of biologists in it, many of whom are brave people who had stood up over war research.  And they wouldn't stand up and testify on the dangers of this research, not even as individuals or just to read Sinsheimer's statement publicly, because they knew that they were going to be in trouble if they took that position.  The reason they would be in trouble was that powerful forces were in motion to develop this technology.  That is happening everywhere, and we have to grab it now and cut it open with a scalpel and really look at it.

AUDIENCE:  Mr. Callahan is here to represent the lay point of view, and he would have us believe that nonscientists should participate in the debate on the issues raised by recombinant DNA research.  I would therefore like to know why it is that Mr. Callahan has refused to endorse the concept of a conference to include largely nonscientific experts on the social and ethical aspects of genetic engineering. This conference has been proposed by a well-known, highly regarded, nonprofit and nonpartisan organization whose name he may wish to mention.

CALLAHAN:  The conference I think you are referring to is one that I felt was not--as initially proposed--going to lead to a good and full discussion of the issues.  I did my best to agitate to see that that was changed.

AUDIENCE:  Your requests were largely met, and I wonder whether the fact that Maxine Singer's husband is an associate of yours in your institution may have had something to do with your feelings on the issue?

CALLAHAN:  I think the conference that you are referring to also got endorsements by other associates of mine, one being Richard Roblin, an original signer of the Berg letter.  If you want to get endorsements of conferences, you got a good mixture, finally.

TADAO JAVITSKY:  I am a senior citizen.  I would like to ask if the great scientists here who are going through this rigamarole and these great discoveries and so on, have they really looked down inside themselves and seen if they are doing what they feel is morally right? Are they looking at the spiritual and the religious dimension?

DANIEL E. KOSHLAND, JR.:  I will answer that one, even though I want to make it clear what I am saying now doesn't prejudge the discussion on recombinant DNA.  I think a lot of us have various positions.  But if I feel that there is a drug that could be developed which has more benefits and will save many more lives than the risks in developing it, I think it is my moral obligation to present that.  That is the business I am in, and I think it is my responsibility to present it to the public.  What if you had mentioned penicillin forty years ago, and somebody had said that this kind of antibiotic is going to change the flora and fauna of your stomach, a certain number of people are going to die from allergic reactions, should we go ahead with this? The scientists could not have guaranteed that it would be a wonder drug that would save millions of lives and wipe out the biggest source of death in this country and the world.  They had to go ahead with some risks.  I think the moral obligation of all of us is to balance the benefits and the risks, and I think there comes a point when the moral obligation is on the other foot also.

If you so exaggerate the risks we would never have an automobile. You would say what is the purpose of having some device that can go slightly faster than the horse and buggy and is going to kill 50,000 people a year a couple of hundred years from now.  All progress can be weighed in benefits and risks, and I think what Dr. Singer was trying to say earlier and what everybody has to weigh is what type of benefits and risk ratio. So it is my moral obligation to say it.

AUDIENCE: I am definitely a nonscientist, but I came here to learn something about DNA. I wanted to learn about not only the risks and the other considerations that one group has largely been discussing, but I want to know about the other side of it. I want to know what the scientists have to say about it. I think the other group has represented itself more so than it should in a democratic process. I would like to get on and learn more about DNA.

AUDIENCE: I sort of grew up with the nuclear bomb as my godfather, and during the course of the last few decades and into the discussions now going on about nuclear safety we have always had the problem that once the genie was out of the bottle, once you guys discovered how you were going to split the atoms and make those reactions, nobody could really control how each individual country or each individual scientist applied that pure research to his or her given ends. If you are quite willing to take the responsibility of the benefits of this, I want to know if you are willing to put your name on the plaque that says our research made it possible for somebody to abuse it. Who is going to take the responsibility for the abuse of the physical research? That is what I want the answer for, and so far I haven't heard anything even approaching that.

JONATHAN BLOOM, Washington, D.C.: I have a degree in mathematics, but I have never studied biology. Actually, I am also in favor of at least a temporary moratorium on this type of research. I would like to know the names of other companies and also any government agencies which indirectly or directly support any types of recombinant DNA research. And I would also like to know what type of research that is, what actually they are going.

HAMBURG: I have the feeling that that really is the next two days, but is there anybody who wants to give a short answer to that?

ERICA THORN, Institute for World Order: You mentioned that there were no strings attached to the funding of this conference, and Maxine Singer urged us both not to restrain the search for knowledge and to recognize the difference between acquired knowledge and applied knowledge. In the same breath she mentioned also that there were not the facilities for the type of research that could go forward. There was not enough money for a full exploration of recombinant DNA research. I wonder where that money is going to come from. Evidently it is not going to come from the places that employ you all. Is it possibly going to come from Abbott Laboratories, Ford Foundation, Smith Kline and French Laboratories, the Upjohn Company? And if so, do you suppose that they are honestly going to want to just acquire this knowledge and keep it to themselves? Are they not motivated specifically to use the knowledge, to apply it for a profit motivation, and that gets us back to the same question. But that is a serious question. Is that not where the money for the research is going to come from?

SINGER: I think perhaps you mis-heard me. I did not at all discuss the funding of this research. In fact, the National Institutes of Health, the National Science Foundation, the Department of Agriculture, to my knowledge, all are funding work in biology which involves the use of recombinant technology. I didn't mention at all any question of funding. I talked about things being slow because of the need to comply with the requirements of the guidelines, not because there was a shortage of money for this research.

THORN: But there was an inadequacy of facilities, yes?

SINGER: Oh, there is an inadequacy of facilities because most institutions do not have facilities which comply with the requirements of the NIH guidelines regarding the definition of physical containment facilities, and not because they don't have the money specifically, but because the facilities have not been built and certified as yet.

## POTENTIAL RISKS OF THE RESEARCH

Erwin Chargaff

*Professor Emeritus of Biochemistry, Columbia University*

One of my most frequent sayings during my long career as a research scientist has been: "Never say no to an experiment." Generations of graduate students must have heard me say that, at one occasion or another, when we were discussing the value of a particular experimental approach. This means that I am very much of an anti-Cartesian. The motto is not: *Cogito, ergo sum*; it is much rather *Sum, ergo cogito*. It is, therefore, imperative for me to explain why I relinquish my old maxim in the case that we are discussing here. There are practical considerations, but there are also--if I may use words that will sound obnoxious or ridiculous in many molecular ears--there are also ethical or even metaphysical considerations.

Scientific research is no longer possible without the complete support by the nation, that is, by the people who pay the taxes taken from them and who breathe the air left to them. The universities are in no position to support innovative research; quite the contrary, a not inconsiderable portion of their income derives from the so-called "overhead," a form of *pourboire* allotted them by the nation.

This shift in the support basis of pure research, which has occurred in my lifetime, means that science has become a political issue, as does everything depending upon the state. This also means that major scientific decisions are no longer exclusively left to the scientists; they

45

are subject to the scrutiny of the elected representatives of the people. This scrutiny is exercised very poorly and inefficiently, since the myth of expertise, which should have no influence on problems affecting the future of all people, has a paralyzing effect on the watchmen who usually look in other directions than the important ones.

People all over the world have become extremely aware of the disastrous deterioration of the environment that industrial and scientific progress has brought them: the contamination of air and water with chemicals and also with radiation--and as a probable consequence the tremendous increase in malignancies--the unsolved problem of the disposal of nuclear waste, being made even more intractable by the sprouting of nuclear reactors, etc.: all this amounting to an interference with the homeostasis of the world, with the evolutionary balance, as the history of civilizations has never experienced before in its long course. Basic science, one of the greatest gifts of the human intellect, must avoid even the appearance of adding to the misery of mankind. I fear that there is a great likelihood that the research on recombinant DNA that is already being performed, and especially the direction in which it is going, will contribute to the impermissible load that our generation has been imposing on the future.

Since, as I have said before, scientific issues have now become political issues, it is not surprising that everybody courageous enough to resist a majority trend will be attacked with all the dirty weapons so plentifully available in the arsenal of politics. This has happened to me and will doubtless happen to all others taking a similar stand. I shall not reciprocate by mentioning names or impugning anybody's motives, which, I am sure, are invariably extremely noble. Taking a historical view, I shall merely point out that minorities are often vindicated by future events, but never before it is too late.

Even now it is probably too late, for I have the impression that what can happen has already begun to happen, although perhaps not yet on the scale at which it will happen in the future. Voluntary moratoria are about as effective as a withdrawal cure in the case of a severe drug addict, although the doctor in charge may benefit.

Before I go on, let me say one thing. I should not dream of asking my congressman whether it is a good idea for me to study the stereochemistry of some sugar synthesis. It is none of his business, just as he does not ask me whether he ought to accept some South Korean needle money. Everyone stays in his own profession. I do not even want him to inquire from me why I do a particular piece of research, although he may be entitled to do so as a watchman over federal expenditures. But it is different if my work impinges upon the health or safety of others. In this case, it is not even sufficient for me to declare my willingness to drink, in public, the elixirs which I have brewed, in order to demonstrate their harmlessness, for I may damage myself as much as I want to with my chemistry, but not one iota of danger to others is permissible. This is my version of the Golden Rule of science.

Thinking of the marvelous series of advertising cartoons that the

*New Yorker* ran a few years ago, you may conclude correctly that I am the only one who does not read the *Philadelphia Evening Bulletin*. I have, however, been reading other literature; and what I saw there, I did not like.

I have been very parsimonious in my own statements on the problem under discussion: only one letter to the editor of *Science* (June 4, 1976). The reason for my reticence is that I am deeply pessimistic about my ability to stop, or even only to slow down, the rush. The juggernaut of scientific majority opinion is much too strong for a few individuals to have any effect. I can only hope that the names of the many workers of the first rank who have assured us that nothing can happen will be remembered when something does happen. I also hope that you will see an essential difference between, say, Kolbe's protests against Kekulé's benzene structure or van't Hoff's tetrahedral carbon model and the few of us who warn against the danger of an irreversible pollution of the biosphere that may have slow but far-reaching consequences.

In the rest of my remarks I shall start from a few facts that I consider, most unfortunately, as accomplished facts.

The decision has been made--and it is an irrevocable decision--to use variants of the obligatory human symbiont *Escherichia coli* as the host. I assume that the number of laboratories observing P1 and P2 contaminant conditions--designated, presumably, by signs saying "Wash your hands when you go home"--will soon go into the hundreds, and that there will be a quite considerable number of P3 and P4 laboratories, all supervised by biohazards committees practicing brotherly love. Therefore, I must conclude that escapes, and sometimes quite massive escapes, will occur at one time or another. I see no way of monitoring them. The recent suggestion by the citizen's committee of Cambridge that the intestinal flora of the laboratory workers should be examined at regular intervals is most praiseworthy, at least for making the whole thing a little more unappetizing. But what will they be looking for? Bugs with a Swedish accent?

I assume that the NIH guidelines in all their inadequacy will eventually, but probably too late, acquire legal status and become enforceable, whether by federal or by state laws. This has already given rise to pettifogging attempts to circumvent even the mild restrictions contained in this document. I see even now papers in which it is explained why certain experiments calling for P3 containment were performed under P2 conditions. It is to be feared that a great deal of rabulistic smartness will go into beating the guidelines.

I assume that the pharmaceutical industry all over the world is already engaged in preparing for massive experimentation and production. They will certainly find ways, even if the guidelines become law, to abolish the ten-liter limit suggested for cultures of reconstructed cells. The industrial applications of the new techniques are undoubtedly promising; but I doubt that the drugs thus produced will be cheaper than they are now.

I assume that equally or even more objectionable experiments not employing recombinant DNA per se, for instance, those making use of some sort of transduction or reconstruction, have been going on all the time; and that these do not even fall under the mild restrictions of the guidelines. As a matter of fact, science is not equipped to restrain or police the sick imagination of a few of its practitioners.

I could go on enumerating the few benefits and the many drawbacks that I can discern in the type of genetic research that has begun a few years ago and that will soon proliferate to an unheard of degree--in small part with the aid of my most reluctant tax money--but this will be done better by others.

I turn, instead, to my last theme. If there existed a Platonic commonwealth, governed by a few sages, how would they have faced the problem we are discussing? They would certainly have said that the genetic inheritance of mankind is its greatest and most indispensable treasure, which must be protected under all circumstances from defilement. With all due respect for the cleverness of molecular prestidigitation they would have declared that the Public Health Service is not the agency to meddle with so all-important an issue, and they would, perhaps, have decreed as follows:

1. All genetic experimentation would be a federal monopoly to be supervised, licensed, and financed through an authority in which representatives of the people would by far outnumber the scientific spokesmen.

2. One, two, or three national laboratories in relatively isolated places would be designated as the only places in which this kind of research could be performed. These laboratories would be charged with carrying out a wide range of preliminary and safety investigations.

3. The most important task, however, would be an intensive and liberally supported search for other less objectionable microbial hosts.

But ours is not a Platonic commonwealth, and I fear we are sliding into an awful mess. In conclusion, I should like to say that anyone affirming immediate disaster is a charlatan, but that anyone denying the possibility of its occurring is an even greater one.

# POTENTIAL BENEFITS OF THE RESEARCH

## Daniel Nathans

*Boury Professor and Director, Department of Microbiology*
*The Johns Hopkins University School of Medicine*

Since the issues raised by the use of recombinant DNA were first brought
to public attention in 1973, a great deal has been written or spoken
about the potential benefits and possible hazards of research with recombinant DNA. After some three years of experience with the method, we
now have a firmer basis for assessing its usefulness, and that is what
I have been asked to do in this initial presentation on the potential
benefits of recombinant DNA research. During this Forum you will hear
from experts in various fields about specific current and planned applications in much greater detail, and of course you will be hearing about
possible hazards. In this overview I shall first make some general
comments about new experimental approaches opened up by recombinant DNA
methods, then examine briefly what kinds of research in biology deserve
to be called beneficial, and finally give some examples of actual or
potential applications of recombinant methods both to basic questions in
biology and to practical problems in agriculture and medicine, particularly the latter, with which I am more familiar.

Nearly everyone who has written about recombinant DNA has recognized
that the ability to clone and amplify segments of DNA from any source
opens up new experimental approaches in biology and medicine. One of
these new approaches can be characterized as analytical, whereby large,
complicated cellular chromosomes are dissected into smaller, homogeneous
segments, each amenable to the types of biochemical and genetic analysis
carried out so successfully with small viral chromosomes over the past
two decades. In short, by this means complex chromosomes can be studied
chemically and biologically in detail, piece by piece. A second approach,
that might be called synthetic genetics, is one in which the reactants are
DNA segments comprising genes and various controlling signals for DNA
replication or transcription, or for protein synthesis. The initial
products are recombinant DNA molecules capable of expressing their genetic
information inside cells. The final products are either proteins encoded
in the transplanted genes or the products of reactions catalyzed by those
proteins. By means of this method any protein whose structural gene has
been cloned could, theoretically at least, be mass-produced. Add to
this a knowledge of the nucleotide sequence of the cloned DNA, i.e., the
order of its subunits, and the ability to create mutations at preselected sites within that sequence, all now feasible, and one has a new and
powerful method for systematic synthesis of a wide variety of protein
analogues. Some applications of these methods I shall come to presently.

The second general comment I wish to make concerns the kinds of research that are beneficial or potentially beneficial. Specifically,
*is fundamental research in biology beneficial*? I raise this question

even here at the National Academy of Sciences because certain commentators on recombinant DNA have questioned the social value of research aimed at answering fundamental biological questions, as opposed to applied research. Quite aside from the intrinsic value of creative attempts to understand as much as we can about ourselves and other living creatures, based on the historical record we can conclude that fundamental research in biology is demonstrably socially useful. In many areas of applied biology and medicine the simple truth is that we know too little to decide what approaches are likely to be fruitful. (A brief, documented presentation of this view is included in the 1976 report of the President's Biomedical Research Panel, and I won't attempt to document it here.) Therefore, I include under potential benefits of recombinant DNA research contributions to understanding basic processes of life as well as practical applications.

What is the potential contribution of research with recombinant DNA to understanding basic biological phenomena? In my opinion, and that of many others, it is considerable. Some of the most important questions that have engaged biologists for a century and more are related to the organization and expression of genetic information in cells of animals and plants. How are genes organized within chromosomes? How is DNA duplicated and passed on from one generation to the next? What are the regulatory signals in DNA and what molecules interact with them? What is the nature of genetic programs for development of an adult from a fertilized egg? How do cells grow and develop into specialized cells and tissues, such as muscle or nerve? How have the structures of genes and chromosomes changed during evolution? These are questions about fundamental phenomena of which we are still largely ignorant, and among the reasons for this ignorance is the complexity of chromosomes of higher animals and plants.

In the case of man, probably hundreds of thousands of genes are encoded in the DNA of every cell. How does recombinant DNA technology help answer these questions? With the analytical methods I spoke of earlier, cloning of appropriate recombinants leads to the isolation of single genes or gene clusters from chromosomal DNA, together with regulatory sequences. An impressive start has already been made in analyzing such cloned DNA segments from fruit flies, frogs, and sea urchins. For example, we now know in some detail how a few gene clusters are organized within a chromosome, including information on the nucleotide sequences of DNA segments between individual genes. And in the case of the fruit fly, *Drosophila*, an object of intensive genetic study for many decades, the distribution of probable regulatory sequences along cellular chromosomes has been mapped by using cloned recombinant DNA segments. Also, an unexpected mobility of genes has been suggested by recent experiments that could be a clue to basic mechanisms of turning genes on and off during normal or abnormal development. Extension of these analyses to other gene clusters and to other animals is underway in a number of laboratories. As a result of experiments of this type, construction of maps of complex chromosomes, including those from man, will be possible, and some general principles of chromosomal organization, evolution, and regulation are likely to emerge.

General methods have also been developed recently for cloning DNA transcripts of cellular messenger RNAs (which are derived from genes) such as that for mammalian hemoglobin and for antibody. In this way a number of preselected genes can be cloned, and in turn this cloned DNA can be used to purify and finally clone larger chromosomal DNA segments that contain specific genes and probable regulatory signals surrounding them. To study the expression of these genes in cell nuclei and the effects they have on cells, it is now feasible to link them to DNA segments from animal viruses and have them propagate in the nuclei of animal cells in culture or become incorporated into cellular chromosomes. In this way it should be possible eventually to identify and characterize specific cellular components that regulate sets of genes in animals and plants, a goal that has eluded many investigators despite years of effort.

The same methods can be used to study many human hereditary disorders, of which there are a large number. For example, in patients with a hemoglobin disorder called thalassemia it is known that there is an imbalance in the synthesis of the $\alpha$- and $\beta$-messenger RNAs of globin and therefore of the $\alpha$- and $\beta$-polypeptides of hemoglobin, often resulting in severe anemia and early death. Since in one of the commonest forms of this disease, $\beta$-thalassemia, the hemoglobin chains are not themselves abnormal, it is thought that there is defective regulation of transcription of the $\beta$-globin genes. To understand this defect it would be essential to isolate these genes and their regulatory signals from normal and from thalassemic individuals, to compare their detailed structures, and to study their expression in the test tube and in normal or thalassemic cells in culture. All of these steps should soon be possible. A good deal would be learned by this type of study about normal as well as abnormal regulation of globin synthesis, both in the adult and during fetal development of the blood-forming cells. I should point out that what is learned about regulation of globin genes could have eventual therapeutic application. Since there are multiple globin genes in a given individual, only some of which are defective in patients with thalassemia or other hemoglobin disorders (such as sickle-cell anemia), it may be possible at some future time, when we understand more about regulation of these genes, to switch off the defective globin gene and turn on a normal one, thus compensating for the defect.

Similar approaches can be applied also to analyze the complex genetic programs that regulate cell growth and division and the abnormal growth that leads to cancer. At present we simply do not understand how the multiplication of human and other animal cells occurs, and how a normal cell changes to a continuously proliferating cancerous one. These are fundamental questions in cancer research, and it is difficult to imagine how we can find meaningful answers to these questions without dissecting the complicated genetic program for cell growth into its component parts.

In my own area of research with tumor viruses, recombinant DNA methods are likely to play an increasingly important role in understanding how viruses cause cancer. During viral tumorigenesis there appear to be two crucial events: first, incorporation of a functioning viral gene into cellular chromosomal DNA; and second, modulation of chromosomal gene activity by the protein product of the viral gene. To help understand these events, one needs to know the structure of incorporated viral genes and which cellular genes or regulatory elements are acted on directly by the viral protein. Recombinant methods could provide a means for isolating and studying the relevant cellular DNA segments.

Now I want to turn to some potential applications of recombinant DNAs to practical problems in biology and medicine. First of all, since recombinant DNAs extend the range of genetic variation of microbes, any industrial microbiological process might be improved by recombining genes and their regulators in new ways, in the simplest instance by transferring genes from a microbe of interest into specially constructed strains of *E. coli*. This technique is not limited to *E. coli*, but at present far more is known about the laboratory strain K-12 of *E. coli* than about any other bacterial cell. Strains can be constructed that will express added genes at high rates and at the same time be defective outside the laboratory. With such recombinant bacteria it has already been possible to multiply yields of enzymes manyfold, and in the future it may be possible to improve yields of valuable fermentation products, or of antibiotics, or to improve the protein content of single-cell-animal feed products now under development.

In regard to other applications in agriculture, there has been discussion among plant biochemists about the possibility of increasing the efficiency of biological nitrogen fixation, one of the limiting factors in crop yields, by transfer of bacterial genes for nitrogen fixation into free-growing soil bacteria, or into plant cells directly. Also discussed is the possible use of recombinant DNA to construct new hybrid plants. These applications are far from my own area of competence, but I gather that a great deal of work will be required to determine whether any of these ideas are practical.

Several medical applications of recombinant DNA methods have been proposed, many of which depend on the production of human proteins in bacteria or other cultured cells. I say "bacteria or other cultured cells" to indicate that since it is possible to infect cultured animal cells with self-propagating recombinant DNA molecules, if bacteria or other microbial cells prove unsuitable it probably would be possible to use more costly animal cells. But what is the likelihood that bacteria can be used to make human proteins? So far we know that the yield of bacterial proteins can be increased manyfold by recombinant methods, as I already indicated, and that certain yeast genes can be expressed in *E. coli*, but to my knowledge, no animal gene has yet been successfully translated into protein in bacterial cells. However, there do not appear to be insurmountable differences in the protein biosynthetic pathways of bacteria and animals. One possibly important difference may be the start signal for protein formation. In the past few years DNA sequences involved in starting and stopping bacterial gene tran-scription and protein synthesis have been precisely identified, and it should be feasible to construct recombinants with active human genes next to appropriate bacterial signals, thus allowing translation of human genes into proteins in bacteria. Whether such proteins would be biologically active may depend on subsequent modification, such as specific enzymatic cleavage, changes in particular amino acids, or addition of sugar residues. In such cases specific methods would have to be worked out to modify individual proteins.

What human proteins have potential medical value? If one were to put this question to medical investigators in various fields, I think the resulting list would be a long one. Among these proteins are those lacking in patients with certain hereditary diseases, diabetics requiring insulin being one of the most prevalent, and hemophiliacs requiring specific clotting factors being a rarer group. Several other diseases are known in which an extracellular protein is deficient.

Another human protein of possibly great therapeutic value is interferon, an antiviral protein with little toxicity that has shown clinical promise against viral diseases and against certain forms of human cancer. Lack of availability of this protein, now produced from suspensions of human white blood cells, has precluded adequate therapeutic trials in patients. However, active messenger RNA for interferon has been partially purified and could serve as a starting point for attempts to clone the interferon gene in bacteria, as has been done for rabbit hemoglobin. There are certainly technical problems due to the small amounts of interferon RNA produced in cells, but with perserverance these problems are likely to be solved.

Other proteins with possible therapeutic as well as experimental value include several immunologically active ones: specific and highly potent antibody molecules, which could be useful in the treatment of infectious diseases caused by drug-resistant organisms; analogues of antibodies that might block steps in allergic reactions; and possibly analogues of those inflammation-producing proteins involved in the tissue damage seen in autoimmune diseases. What I am suggesting is that some time in the future recombinant DNA techniques may provide, for the first time, rare, biologically active human proteins or their analogues in sufficient quantity to study their mode of action as well as consider their use as therapeutic agents.

Another potential application I want to mention is the production of vaccines. Virus vaccines are now produced from viruses grown in cultured cells or in living animals, for example, chick embryos in the case of influenza vaccine. At times vaccines become contaminated with other, often unrecognized viruses derived from the cells or embryos in which the vaccine virus is prepared, or with cellular components that cause allergic reactions in certain recipients. There would be obvious advantages, aside from reduced cost, to preparing immunizing viral proteins in bacteria, starting with the one or two appropriate viral genes linked to bacterial plasmids. A similar argument applies to certain bacterial vaccines in those instances where genes known to code for pathogenic proteins have been identified. A notable, and in my mind a promising, example would be a vaccine against cholera, still a widespread disease in the developing world. To determine whether this general approach is practical would obviously require extensive experimentation and testing.

Finally, I want to say a few words about gene therapy, a potential application of recombinant DNA research often discussed. The notion has a certain simplicity: since many hereditary diseases are due to a single defective gene, addition of a normal copy of that gene will restore functional gene product. In my opinion, there are many theoretical, practical, and social problems to be solved in this area, and we are a long way from any attempts at gene therapy in patients. Yet there are experiments with cells in culture and with mice that bear on this possibility. First, cultured animal cells transfected with segments of recombinant viral DNA or with segments of normal chromosomes can acquire heritable, functioning genes from the donor DNA. Second, it has been shown that when mouse embryos were infected with viral DNA and reimplanted in the uterus of a foster mother, and the resulting adult animals examined, many of their cells contained viral DNA sequences. It therefore appears possible to introduce persistent genes into individual cells in culture or into many cells in a living animal. These procedures coupled to recombinant DNA methods could clearly be of considerable importance in understanding the effects of normal or defective genes in cells and in living animals. Whether they bring the practical possibility of gene therapy in human disease any closer is less certain.

To sum up, recombinant DNA methods have opened up new approaches to problems in fundamental and applied biology. Nearly everyone agrees that the methodology is a major technical advance. In my opinion, it is likely to have at least as great an impact in biology and medicine, and on human welfare, as the development of cell culture methods, a technique that revolutionized the study and medical usefulness of cells and of viruses.

I have cited in this overview several examples of actual or potential applications of recombinant methods in basic research, in agriculture, and particularly in medicine. Some of these applications have progressed to the point where substantial benefits are already evident, as in the analysis of complex chromosomes and mass production of certain enzymes. Other applications appear promising for the near future, particularly for understanding hereditary and infectious diseases and cancer, for the production of useful human proteins, and for the development of new vaccines; and still other applications, such as the production of new hybrid plants and gene therapy for patients, are speculative and more distant.

With a method so broad in its implications we obviously cannot foresee all its possible uses. As the technology develops, as understanding of genetic mechanisms increases, and as more scientists with a variety of interests begin to use recombinant DNA methods, additional applications that don't occur to any of us today will surely follow.

# DISCUSSION

GEORGE WALD, Professor of Biology, Harvard University:  It seems to me
that the entire Forum may be absorbed with discussing the safety or
lack of safety of this kind of research.  I had hoped in coming here
that we would tackle the very much bigger question of not how to do
this research safely, but whether to do it at all.

I have heard the issue raised by Maxine Singer and others of sup-
pressing free scientific inquiry.  Jonathan King has already spoken
of this.  No one I know is trying to suppress the inquiry.  All of
us, indeed, are asking the same questions and would like to see them
answered.  The objection here is to this specific technology for
attempting to answer those questions.

I have spent my life in science, and my understanding of the
enterprise I was engaged in was trying to understand nature.  It did
not involve the manipulation and deformation of nature.  I keep, as
I hear this kind of discussion, thinking of that sad major in the
American Army in South Vietnam, who said of a city that had been
bombed out by the Americans, "We had to destroy it in order to save
it."  The argument here is we have to twist it out of the whole
natural order of life in order to understand it better.  We are all
asking the same questions and would like to see them answered.  It is
not as though there were not alternative methods for answering those
questions already available, and I trust that the future will present
us with more.  So for example, there is an enormous amount that can
be done working within single species or species that regularly
exchange genetic information in nature, such as all the flora of
the human bowel.  If one has to go on with *E. coli*, it regularly
exchanges genetic information with other members of that flora, and
if one restricted oneself to that kind of experiment, I think that
one could indeed approach and answer many of these questions.

A second alternative method is what Khorana and his group did in
synthesizing both the gene and its control mechanism, and then with
complete control of the situation inserting this back in that very
same species and strain that lacked the gene that they had synthesized.

Now, if one thinks in these directions, it seems to me that the
whole argument for recombinant DNA research reduces to a plea for
convenience and speed, which I don't think are the essential considerations in this situation.  What Khorana did was very laborious.  It
took nine years for a rather large group to perform that synthesis.
But having done it he knew exactly what he was doing, which is true
of very few recombinant DNA investigations of which I have heard.

You know, one talks very familiarly of isolating a molecule and
inserting this molecule.  What one generally isolates is a block of
genes of which one has characterized one or two.  There is a lot
of unknown material that goes along with them.

So I hope that this Forum doesn't end as it has begun, with the
assumption that this is a proper line of investigation and all we have

to be concerned about is its safety. I think this much larger issue should take precedence over that, and I see no indication that this meeting intends to approach it.

STANLEY N. COHEN, Professor of Medicine, Stanford University Medical Center: Some of the comments that Professor Chargaff and others have made have implied that evolution has remained in some primeval state for millions of years, previously untouched by man, and that recombinant DNA now threatens natural evolution. However, there is no evidence at all that evolution is presently under delicate control by nature. Man has continually and for many, many years altered the progress of evolution in various ways. Initially, the domestication of animals and the cultivation of crops by man gave selective advantages to certain species and enabled their propagation in new environments. Later, man constructed hybrid animals and hybrid plants, and in this way created new species that did not exist naturally. Man has continued to alter evolution in other ways. By the use of mass immunization programs for viral disease man has altered the evolution of viruses, and of course by the development of antibiotics for the treatment of infectious diseases man has altered the progress of bacterial evolution.

Yet, Dr. Chargaff still asks whether we have a right to "alter the evolutionary wisdom of millions of years." I would like to point out that this so-called evolutionary "wisdom" has given us bubonic plague, and smallpox, and yellow fever, and typhoid, and diabetes and cancer. The search for and the use of virtually all biological and medical knowledge represents a continual and intentional assault on what Dr. Chargaff considers to be evolutionary wisdom. Most post-Darwin biologists believe that there is no wisdom in evolution, only chance occurrences. Do we really desire to glorify chance evolutionary occurrences as "wisdom" and to accept without protest or countervening action the diseases and plagues that such "wisdom" has bestowed on mankind? I would suspect that most of us are not prepared to simply endure whatever nature may have in store. Thus, science continues to search for new ways to influence the "wisdom" of evolution.

I would like to end with one question of Professor Chargaff, and that concerns his initial remark that he never says no to an experiment. I have, in the past few weeks, looked through some of his writings of the early 1970s in which he deplores the loss of innocence that science has suffered in his lifetime. In 1971, long before the advent of recombinant DNA, he notes, "It would seem to me that man cannot live without mysteries. One could say the great biologists worked in the very light of darkness." That, sir, does not seem to me to be consistent with your claim of not saying no to an experiment. This statement and others suggest that you have for some years been deploring the search by man for scientific knowledge to dispel the mysteries of life.

CHARGAFF:  I saw Dr. Cohen's paper in *Science*, and I realize he gave a long list of complaints against God, the plague, and all that.  I, in all humbleness, in all humility, am willing to bear God's scourges. But do I have to add Dr. Cohen's scourges to them?

What I have warned against continuously is that we are putting an additional, fearful load on our biosphere.  We can't change our past history, and we are all human.  But do we have to do unnecessary things without knowing what we are doing?

As to Dr. Cohen reading in my collected works, I am glad to have found one reader.

BERNARD D. DAVIS, Professor of Bacterial Physiology, Harvard Medical School:  At the end Dr. Chargaff recommended certain restrictions that he would like to see, and he said something about all genetic research.

CHARGAFF:  No, I meant all the recombinant DNA research.

DAVIS:  You would not include transduction?

CHARGAFF:  I love genetics.  I have nothing against the breeding of pure-breds.  Some of my best friends are horses.

DAVIS:  Well, in other words you don't object to the isolation of bacterial mutants?

CHARGAFF:  Look here, Bernie, no one really asks me.

DAVIS:  We slid from one thing to another very easily.  You introduced transduction as something almost as bad as recombination.  We have had thirty or forty years of experience.  The fantastic development of molecular genetics has come from mutations, from gene transfers within bacteria, and from the ability to identify and map a gene.  Khorana can isolate and synthesize a gene because it has been identified.  Nobody can identify a gene on a chromosome, with very rare exceptions, unless he has a mutant form to recognize.  So we are constantly manipulating, perturbing, disturbing nature.  And while I admire Dr. Wald as being able to do experiments that don't perturb nature, most of us to some degree have to perturb nature.

CHARGAFF:  The trouble is, you see, the Virgin giving birth to a little baby and saying that it is just a tiny baby, not very big.  I am not quite sure where you draw the line, and I am not able at this time really to define my philosophy of science.  But I am sure that if you try to think quietly, instead of talking all the time, you will perhaps see what I am driving at and where the difference does lie.

# DAY II

---

ALEXANDER RICH
Sedgwick Professor of Biophysics
Massachusetts Institute of Technology
*Cochairman*

# PRIORITIES FOR DAY II

Alexander Rich

The Academy Forum was designed to examine controversial issues, to
discuss all aspects of them in an open manner, and to reach maximum
public involvement.  What we really are trying to do is bridge the gap
between science and scientists and people and to do this on issues that
are of public interest.

As a component of this process, we have assembled a Panel for Inquiry,
a broadly based group of people who will illuminate many aspects of the
discussion, pose critical questions, and explore dimensions of the problem that we did not incorporate into the program explicitly.

As another component of the Forum, we have this time introduced
workshops.  The workshops allow us to discuss in a more intensive
fashion issues that were not in the formal program.  We are hoping that
people will add suggestions about workshops that can be formed.

Yesterday's discussion was a broad one, covering a variety of fundamental issues, some of which were raised by the speakers, and some raised
by the audience.  This is the essence of the Forum process: a dialogue
between scientists, speakers in the Forum, and people in the audience
who think about the subject and bring out aspects of it that were not
covered in the formal presentations.

In organizing this Forum we adopted a new mechanism of exploring the
subject, not by covering all aspects of it but rather by focusing on a
few discrete areas.  We call this case analysis.  Its virtue is that it
allows us to describe in great detail certain issues and to develop
examples that illustrate the broader scope of the problem.  We cannot
discuss everything at once.  So now we will focus on a few issues and
explore them in depth.

The first case analysis concerns the mapping of the mammalian genome.
This is an enormous scientific enterprise:  What does this mean in terms
of possible benefits and possible risks?

61

## POTENTIAL BENEFITS

Paul Berg*

*Willson Professor of Biochemistry, Stanford University Medical Center*

Extraordinary advances during the past forty years have provided a deeper understanding of the chemical nature of genes and how they work. Instead of a conceptualization, a gene is a specific linear segment in a giant molecule of deoxyribonucleic acid (DNA). Each gene or segment of DNA is distinguished from another by the order of four different chemical subunits--adenine, thymine, guanine, and cystosine--that comprise its two intertwined chains.

We know that genes perform two functions: they direct their own reproduction, or replication, during cellular multiplication, and they also direct the synthesis of proteins--the order of the four subunits in each gene segment of DNA defining the biologic activity of a particular protein.

Our present generalized view of heredity is that the properties and behavior of living things are transmitted from one generation to the next during replication of the genes and that an organism's phenotype,

*I am deeply indebted to David S. Hogness and his colleagues R. Lifton and R. W. Karp for allowing me to quote and use their findings before they were published, and to the many people who listened to me as I thought through the ideas in this paper.

i.e., its metabolic capabilities, physical endowments and appearance, and even the susceptibility to infection, heart disease, and cancer, are consequences of the battery of structural, catalytic, and regulatory proteins that result from the expression of its genes.

Much of what is known about the molecular details of gene structure and function has been gathered in studies with simple organisms, principally *E. coli* and several viruses that infect it. These organisms became the favorites because they reproduce rapidly and are readily grown under controllable laboratory conditions. But more importantly, perhaps, their relatively small chromosomes can easily be manipulated by genetic means, and this made it possible to map the location and arrangement of genes on the *E. coli* and viral chromosomes. This achievement was crucial for understanding how the genes of these organisms are expressed and regulated.

The genetic map of *E. coli* may be represented as a circle because its chromosome is actually a single, circular DNA molecule. About 650 out of an estimated total of 3,000-4,000 of the organism's genes have been assigned specific chromosomal locations, indicated by the numbered coordinates on the inside of the circle (Figure 7). Each genetic function is indicated by three letters on the inner perimeter, and individual genes or clusters are arrayed on the outer perimeter. Examining a specific region in more detail (Figure 8) reveals that the map provides considerable information, e.g., it shows how some functionally related genes are clustered (the genes K, T, E, O in the gal group at map coordinate 16.7) and whether genes are read clockwise or counterclockwise (see, for example, the genes AB, PAB, and KTEO in the expanded region between map coordinates 16 to 17). In some regions of the chromosome the resolution of the map permits us to identify genes that regulate expression rather than code for proteins (Figure 9). One such example is a set of genes that govern the ability of *E. coli* to metabolize the milk sugar lactose (the cluster of genes labeled lac-AYZOP). Of particular interest is that genes Z, Y, and A in this cluster specify proteins (enzymes) that permit lactose to generate energy for the cells. Genes O and P control whether proteins Z, Y, and A are made. Even greater resolution has been achieved in this region, since the exact order of DNA subunits in the two control sites, O and P, as well as in the beginning of the Z gene, can be specified (Figure 10). Knowing how the control and protein-coding genes are organized, the subunit sequence of that DNA segment, and the way regulatory proteins bind to this region has provided profound insight into how the production of the Z, Y, and A proteins is modulated in response to the extracellular environment. Similar studies with several other regulated sets of genes in *E. coli* have provided a wealth of solutions to the problem of gene arrangement and expression in this bacterium. These models now guide our search for answers about expression and regulation in the chromosomes and genes of higher organisms.

Our knowledge of the chromosome of *E. coli* contrasts sharply with our ignorance about the molecular anatomy of the human and mammalian genomes. As of October 1975, less than 150 human genes had been

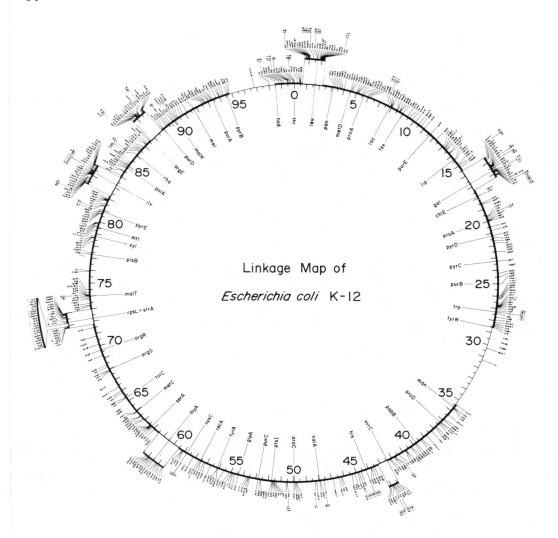

Linkage Map of

*Escherichia coli* K-12

FIGURE 7  Genetic map of *Escherichia coli* K-12.  This representation and the enlarged sections shown in Figures 8 and 9 are from B. J.
Bachman, K. B. Low, and A. L. Taylor (Bacteriol. Rev. 40:116, 1976).

assigned locations on one or another of the 23 chromosome pairs (Figure
11).  About 20 genes have been mapped to regions in human chromosome 1
and about 10 genes to the second chromosome.  But no two human genes
can be located within even two *E. coli* chromosome lengths of each other.
Thus, there could be 5,000-10,000 unknown genes between any two mapped
human genes!  Although research in this field is booming, the acquisition

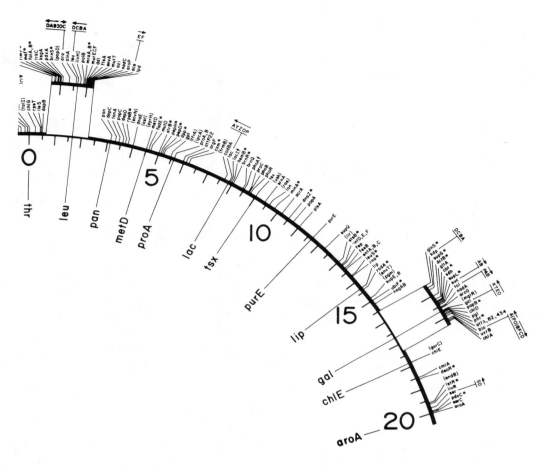

FIGURE 8   An enlargement of the *E. coli* genetic map between coordinates 0 and 20 (see Figure 1).

of results is slow, extremely costly, and very unlikely to provide the level of resolution achieved with *E. coli* genome.  Only in the case of *Drosophila melanogaster,* a fruit fly, has the genome map even approached the resolution achieved with *E. coli.*  About 1,500 *Drosophila* genes have been located on its four chromosomes.  But here, too, it's unlikely that genetic means could provide the level of resolution achieved with *E. coli.*

Deciphering the molecular details of the chromosome of *E. coli* (as well as of several viral genomes) would not have been accomplished without the ability to isolate specific regions of its DNA in pure form and in large enough quantities for analysis.  This was achieved using

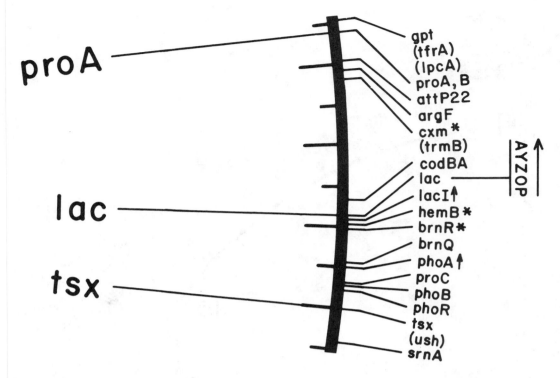

FIGURE 9  A detailed representation of the *E. coli* genetic map between coordinates 6 and 9 (see Figures 7 and 8).

nature's mechanism for producing recombinant DNAs. Sparing the details, bacterial recombination systems can translocate specific gene clusters from the bacterial chromosome to viral and plasmid DNA molecules for selective amplification and purification. Virtually every region of the bacterial chromosome can be manipulated in this way; therefore, the task of solving the molecular structure of these segments becomes straightforward, albeit tedious.

But there are no naturally occurring viruses or plasmids that pick up genes from mammalian or human chromosomes and multiply them selectively. Without such a trick it would be a prohibitively expensive and virtually hopeless task to isolate, by physical means, specific genes or unique segments of DNA directly from the entire DNA complement of a mouse, or rat or man.

Let me illustrate the magnitude of the problem with some simple numbers. The genome of *E. coli* is comprised of a *single* DNA molecule with 4 million subunits. The human genome with twenty-three different DNA molecules, one from each of the twenty-three chromosomal pairs, has 4 billion subunits, i.e., the human genome is a thousand times larger and

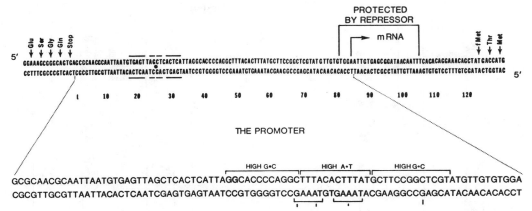

FIGURE 10  The detailed nucleotide sequence of the regulatory region
(P-O) for the lac operon.  The promoter (P) is the region at which
the enzyme RNA polymerase binds and initiates mRNA synthesis for the
production of the Z protein (β-galactosidase).  The repressor binds to
the operator (O) to prevent mRNA synthesis.  The sequence is from R. C.
Dickson, J. Abelson, W. M. Barnes, and W. S. Reznikoff (Science 187:27,
1975), to which the reader is referred for further details.

consequently a thousand times more complex than that of *E. coli*.  It is
more than a million times larger than the chromosome of the bacterial
virus whose entire subunit sequence was recently solved.  Even human
chromosome 1 contains sixty times more DNA than the entire *E. coli*
complement!

Unfortunately, large DNAs are very fragile and easily broken during
isolation.  Consequently, without special precautions, only bits and
pieces of DNA, a few genes in length, are obtained during the isolation
of mammalian DNA.  Obviously the complexity of the mixture of DNA segments, obtained from such genomes, is far greater than those recovered
from *E. coli* or small chromosomes.

But perhaps another example would be more illustrative and impressive.  Consider the result of passing ten copies of this manuscript
through a paper-shredding machine to produce random 1 × 3 inch strips;
now, imagine performing the same operation with ten copies of last Sunday's *New York Times* newspaper.  Most of you would see, intuitively,
that it would be far easier to reconstruct a particular sentence of my
manuscript from its strips than to reconstruct a particular story in
the *Times* from its strips.  What if an urgent decision or national
policy depended upon the information contained in that *Times* article?
And what if only semiliterate persons were put to the task of recon-
structing that article and translating its message?

Scientists, confronted with the incredibly complex mixture of DNA
fragments obtained from human chromosomes, are equally handicapped and

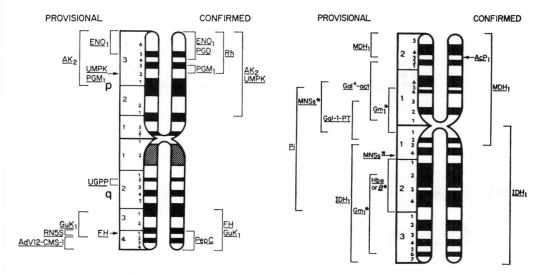

FIGURE 11  A representation of human chromosome 1 (left) and 2 (right) showing the generally accepted pattern of Giemsa-stained banding and the geographic divisions of each arm of the two chromosomes.  Genetic markers are indicated by letter designations, and their appropriate chromosomal locations are indicated by the brackets.  The figure was constructed from the report of the Committee on the Genetic Constitutions of Chromosomes 1 and 2 presented to the Conference on Human Gene Mapping held in Baltimore during October 1975 and published by S. Karger (Basel, 1976).  That volume also provides a more detailed presentation of the chromosomal mapping of other human genes.

frustrated.  For they know that if the complex assortment of DNA fragments could be reassembled in the proper order and the genetic signals deciphered, it would provide a new vision of our genome's structure and have profound significance for improving human health.

With the recombinant DNA method, discrete DNA segments can be isolated in pure form and virtually unlimited quantities from the hopelessly complex mixtures obtained from cells.  Molecular cloning provides the means to reconstruct extended regions of chromosomes and eventually the entire genome of any organism from the bits and pieces of its DNA. This opens the way for analysis of the detailed molecular structure, e.g., the subunit sequence, of individual genes and extended regions of mammalian chromosomes.

These are not speculations.  They are realistic estimates based on the impressive analyses that have already been made with chromosomal DNA segments from several simple higher organisms using recombinant DNA methods.  Astoundingly, the structure of various parts of the chromosomes of a fruit fly (*Drosophila melanogaster*), a toad (*Xenopus laevis*), and sea urchins are now known to the same molecular resolution as the *E. coli* genome.

Although I cannot review each of these contributions in the space allotted to me, perhaps I can illustrate how the experiments can and are being done and summarize what has been learned so far.

Cloned (pure) segments of virtually any chromosome can be obtained by chemically joining random, or specifically cleaved, fragments of their DNA to a vector DNA molecule (Figure 12). The vector DNA, a plasmid (or bacterial virus chromosome), and a segment of foreign DNA are joined in the test tube and then introduced into a special strain of *E. coli* K-12. Conditions can be selected so that only those bacteria or viruses that acquire the recombinant DNA grow or are detected. Because no more than a single recombinant DNA molecule is taken up by the bacterial cell, all the descendants of that transformed bacterium will contain only one particular foreign DNA segment.

It is as if an army of people filed by the collection of *New York Times* shreds and took a single strip to have it copied (Xeroxed) a billion times; that process would clone and amplify discrete segments of the newspaper.

There are a number of different ways to identify cloned mammalian DNA segments. Some segments can be recognized because they contain genes for well-known cellular products (tRNAs, rRNAs), some have genes that code for particularly common messenger RNAs, and some are readily detected because they are unusually abundant in the genome. Whatever

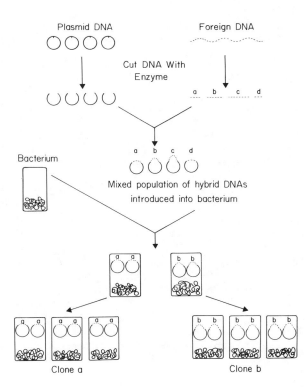

FIGURE 12  A schematic representation of the procedure for molecular cloning using plasmids and bacteria.

way they are detected, the size and chemical features of each DNA segment can be analyzed and catalogued.

But this is only the beginning. Each cloned DNA segment can be cleaved further and the smaller fragments reintroduced into new bacterial cells to obtain cloned subsegments for more detailed studies. Or, cloned segments can be used as probes, to examine other larger segments in the collection for shared and neighboring regions. In effect, one "boot straps" oneself along to reconstruct more extended regions in the chromosomal DNA and to identify the genetic functions they determine.

Once again our analogy with the shredded *New York Times* provides a simple model. Each of you could identify strips coming from the cartoons, advertisements, classified sections, and, probably, the stock market listings. Then it would be straightforward to search the shred library for strips with overlapping information and to assemble the relevant ones into portions or even an entire page.

But how can the cloned DNA segments be mapped to chromosomal locations? David Hogness and his colleagues have already done this with one organism's DNA. A radioactively labeled copy of a cloned *Drosophila* DNA segment will, under proper conditions, be bound to region(s) of the chromosome that contain that segment; the position of these bound segments can be visualized with photographic emulsions that become blackened where radioactive emissions are localizations.

Applying this technique to the giant chromosome of *Drosophila,* Hogness' group has mapped several of the cloned *Drosophila* DNA segments to specific chromosomal sites. In one instance the photographs show that the isolated DNA segment occurs at many regions on all four chromosomes (Figure 13). It is significant that some cloned DNA segments showing such widespread dispersion throughout the genome, contain elements within the segment that occur in only a subset of the chromosomal sites. Does this mean that the same gene might occur in different contexts and be regulated differently at each?

Another cloned DNA segment is found, using the same technique, to occur at only one site on a particular chromosome (Figure 14). In this instance the cloned DNA segment is tandemly repeated at that single chromosomal location. Presently, this technique is applicable only to the giant chromosomes of *Drosophila*, and a way is needed to extend this capability to mammalian and human genomes.

The structural analysis of these simple animal chromosomes has already transformed our views about the physical and genetic organization of their genes. We know that some genes occur as tandemly repeated, multigene families. For example, the genes that code for the two kinds of RNA molecules used in the cell's protein synthesizing machinery (18 and 28$S$ rRNA) occur in two types of such families (Figure 15). The arrangement shown at the top contains repeating segments with two genes, 18 and 28$S$ rDNA segments, separated by short spacer regions. But, unexpectedly, another tandemly repeated rDNA family (shown below) contains an insertion of extraneous DNA within the 28$S$ gene. Although the

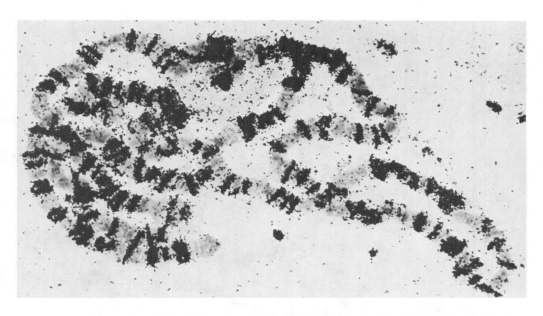

FIGURE 13 *In situ* hybridization and radioautographic localization of a particular cloned *Drosophila melanogaster* DNA segment in the fly's polytene chromosomes. The photograph will appear in *Cell* (1977) in a paper entitled "*In situ* hybridization of [3]H-labeled cRNA to pDm27," by D. J. Finnigan, G. M. Rubin, J. Bower, and D. S. Hogness.

significance of the novel insertion into a gene is still unclear, its implications for control of gene expression are tantalizing, indeed.

Another exciting development is the isolation and structural analysis of a tandemly repeated, multigene family, responsible for the production of five proteins that are associated with the DNA in chromosomes. In sea urchin chromosomes the five genes for the histone proteins occur as a cluster which is repeated, in tandem, 400-500 times (Figure 16, top). The order of the genes and spacers and their molecular size have also been accurately determined. There is also evidence that the entire set of histone genes is read in the same direction--right to left.

However, *Drosophila* chromosomes have a different organization of their histone genes (Figure 16, bottom). Although they too are clustered and tandemly repeated, the gene order is different; but, more significantly, several of the genes are read in one direction and others in the opposite direction. This implies the existence of novel and sophisticated control sites regulating the expression of these genes; moreover, the chemical structure of these control elements is now amenable to scrutiny.

These and other findings that I cannot recount here show that very promising advances have been made during the two to three years that

72

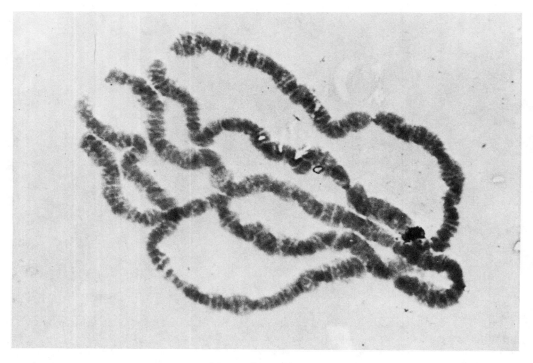

FIGURE 14  *In situ* hybridization and radioautographic localization of a cloned *D. melanogaster* histone gene cluster in the polytene chromosome. This photograph is from R. W. Karp and D. S. Hogness (Fed. Proc. 35:1623, 1976).

FIGURE 15  Two different arrangements of rDNA clusters in *D. melanogaster*.  The data used to construct the figure are from D. M. Glover and D.

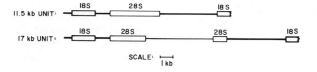

S. Hogness (Cell 10:167, 1977) and R. L. White and D. S. Hogness (Cell 10:177, 1977).

recombinant DNA methods have been applied to the study of chromosome organization.  Now there is an exhilarating optimism that a formerly intractable problem is manageable.  It is no longer a pious dream to expect that the structure of mammalian genome will be known to the same resolution as *E. coli* chromosome.

So far recombinant DNA methods have been applied almost exclusively to the chromosomes of lower organisms.  In fact, until recently, such

FIGURE 16  Histone gene arrangement.  (Top) The organization of
a single repeat unit of histone
genes in a cloned DNA segment
from sea urchin chromosomes;
the arrows indicate the direction of transcription of the
particular nRNA coded by that
region.  This representation
was adapted from data published
by L. Kedes and his colleagues
(Cell 9:147, 1976).  (Bottom)

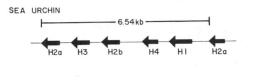

The organization of a single re-
peat unit of histone genes in a cloned DNA segment from *D. melanogaster*
chromosomes.  The parentheses indicate the tentative identification
of those mRNAs, and the arrows indicate the directions of transcription
of the particular mRNA coded by that region.  This representation is
from R. W. Karp and D. S. Hogness (Fed. Proc. 35:1623, 1976) and R.
Lifton, M. L. Goldberg, and D. S. Hogness (unpublished).

experiments with mammalian and particularly human chromosomes were forbidden wherever the NIH guidelines are applicable.  But our goal is
to understand the human genome.  For just as the present-day practice
of medicine is impossible without a knowledge of human anatomy and
physiology, dealing with disease in the future will require a detailed
understanding of the molecular anatomy and physiology of the human
genome.

The riddles of human genetic chemistry were not born with the advent of recombinant DNA methods.  The questions being attacked now have
preoccupied biologists and physicians for more than fifty years.  Moreover, the search for answers to these problems has been actively promoted by the Congress and each administration ever since the end of
World War II.  Literally billions of dollars have been spent training
scientists, building laboratories, and supporting research in molecular
and cell biology to better understand the structure and workings of the
mammalian genome.  With the recombinant DNA breakthrough that task has
been greatly simplified.  Consequently, the perennial questions about
normal cellular growth, development and differentiation, and the basic
causes of genetic ills can be more confidently attacked.  In my judgment the most important practical benefit from recombinant DNA research
will be the knowledge acquired about the detailed structure of mammalian
and human genes and the chromosomes they comprise.  Moreover, I believe
that that knowledge will not only have profound influences on our view
of man's nature but it will also make the diagnosis, prevention, and
cure of disease more rational and effective.

74

## POTENTIAL RISKS

Robert L. Sinsheimer

*Chairman, Division of Biology, California Institute of Technology*

I regret to find myself in what is listed as an adversary position, although I think of it more as an orthogonal position, to Paul Berg, for whom I have the greatest respect as a scientist and a person. In principle one certainly can have no objection to the project of mapping the eukaryotic chromosome any more than one might have to a project of mapping the Sierras. The dispute is perhaps about the technique to be used. One might object to mapping the Sierras if the technique were to move them mountain by mountain to Long Island so that each could be individually measured.

The existence of an intellectual controversy is an indication of uncertainty, of a lack of knowledge that restricts our ability to make intelligent prediction of the consequences of our actions. The controversy may reflect factual uncertainty concerning the nature of the substances or the organisms involved and the general principles that control their interactions. The controversy may reflect human uncertainty concerning the predictability of human actions and the limits of rationality, or the controversy may reflect moral uncertainty concerning the virtues of differing basic value sets.

All of these sources of uncertainty can be seen in the controversy over recombinant DNA. The magnitude of our uncertainty reflects the magnitude of the scientific advance that these new techniques make possible. I would suggest that science has not taken so large a step into the unknown since Rutherford began to split atoms. The recombiners may take comfort in this analogy, for Rutherford's experiments were not in themselves disastrous. He did not, in his ignorance, ignite a consuming chain reaction. In a historical sense, of course, he did, if we include the subsequent three decades of physics in the chain, but I would not be so deterministic.

But, one may ask, will we always be so fortunate in our explorations? Will nature always be so benign and so resilient to our interventions? Are there really no evolutionary booby traps for unwary species? The recombinant DNA technology brings us at one bound into a new domain with great potentials both for good and for harm--and all shrouded by our current ignorance. What are the factual uncertainties that may mask significant hazard and thus pose risks to the unwary?

One large cluster of uncertainties stems from the use of strains of *Escherichia coli* as hosts for much of this research. This organism is known to live in an intimate relationship with man and other animals. It is argued that the K-12 strain employed in this research is not robust and is unlikely to colonize the human bowel.[1,2] The validity of this claim for persons on antibiotics or persons suffering from various debilitating ailments, or human infants or other animal species, is itself uncertain.

It is proposed that we will breed mortality into the E. coli strains to be used in the "more dangerous" experiments.[3] The effectiveness of such breeding in a variety of ecological circumstances remains to be demonstrated. Our ability to define the more dangerous is arguable.

More broadly, the use of E. coli as a host organism points out that we are in considerable measure ignorant of the factors governing the ecology of the bowel. Intricate microbe-microbe and microbe-host interactions involving cross-feedings of vitamins, amino acids, lactate, branched-chain acids, heme, and even hydrogen are known,[4] but no one would pretend we had a full understanding of this microcosm or could predict the consequences of the introduction of novel organisms with novel capabilities.

We are largely ignorant of the normal role of the bowel flora in human nutrition, as in the production of vitamins, or in carcinogenesis, as in the controversy over the importance of bulk fiber in the diet. We are largely ignorant of the effects of plasmids, prophages, et cetera on the fitness of bacteria for bowel survival.[5] We now recognize that many instances of intestinal disorder acquired, for instance, in travel in foreign lands are the consequence of toxigenic plasmids, apparently endemic to the E. coli strains of these countries.[6,7] But we are ignorant of the details of their pathology or the factors governing the fitness of such strains.

We are ignorant of the ecology of E. coli outside the bowel, of the factors that determine its capacity to invade the intestinal wall, to sustain systemic or urinary infection, or to colonize the nasopharynx. In the environment the persistence of coliforms in uncontaminated habitats is still a matter of dispute.[8] Nor is it known what factor or factors, such as bicarbonate, may limit the growth of E. coli in other natural settings and how these limits might be affected in novel organisms.

We are in large measure ignorant of the range and frequency of gene transfer throughout the prokaryotic world. E. coli is known to be capable of genetic exchange with some forty other bacterial species, but we have little knowledge from which to estimate the rates at which novel gene constructions might spread throughout the prokaryotic world under various conditions.

We are ignorant of many aspects of the complex microbiological equilibria that truly underlie and maintain the entire world of life in its present form--quasi-equilibria that affect the bacteria that degrade our wastes and replenish the planetary nitrogen and carbon dioxide that generate our soil and cleanse our waters--and again we can therefore hardly estimate the consequence, short term or long term, of the introduction of novel microbial forms.

We are grossly ignorant of the structural gene content of the eukaryotic genomes that were introduced so blithely into this E. coli. How can we predict the consequence of the interactions of unknown gene products with the numerous macromolecules and metabolites of the *E. coli* organism? The eukaryotic gene products themselves, whether they are the result of faithful or partial transcription and translation, might be toxic to a host. Because we do not know that an organism produces the toxin does not mean it may not have a gene for such. And more generally, what is a toxin?

Products analogous to human hormones, to peptide growth factors or neurotransmitters or releasing factors--consider, for example, a small peptide with insulinlike activity--could have grievous toxic effects.

The DNAs introduced in these strains are in no sense random sequences of nucleotides. They have been, most often, selected by nature to code for proteins that achieve a function, very often a catalysis. The action of such proteins upon indigenous components of *E. coli* might split off polypeptides with unfortunate sequences or might convert normal metabolites into undesirable products--for example, converting amino acids into catecholamines with synaptic functions.

We are ignorant of the factors affecting the penetration of the intestinal epithelium. Some small proteins, such as insulin, can be transported across the intestinal epithelium with an efficiency of a few percent.[9] At the same time, very large proteins, such as the 150,000-molecular-weight botulinus toxin, appear to penetrate relatively readily.[10]

With an estimated $10^{13}$ to $10^{14}$ microorganisms in the human bowel, the production of quantities of insulinlike activity or adrenalinlike activity comparable to the normal human daily syntheses (1 milligram per day of insulin, 250 to 300 micrograms per day of adrenalin) is quite possible, even allowing for limited transepithelial transport.

With regard to novel metabolites one should remember that we are as yet largely ignorant of the etiology of cancer. Does anyone imagine that the roster of carcinogens or mutagens has been completed? We are ignorant of the nature and mode of transmission of slow viruses. Could their ingredients lurk in these random bits of genome we now juggle? Ailments whose symptoms are long delayed are, of course, the most pernicious, for their causative agents could become widespread during the incubation period. We remain largely ignorant of the factors that restrict the spread of viral species among different hosts. It is indeed fortunate that the microbial sea in which we are immersed is not to our knowledge a reservoir for human viral disease. Does this reflect differences in the potential for viral gene expression? Do prokaryotes use different promoters for RNA transcription, different recognition signals for ribosome translation?[11-14] The transferred fragments of DNA must surely carry such promoters and recognition signals. Are the initiation regions for DNA synthesis different in prokaryotes and eukaryotes? A very appreciable proportion of the transferred fragments of eukaryotic DNA will carry the sequence for initiation of eukaryotic DNA synthesis, since these are to be found in every thirty to forty microns of eukaryotic DNA.[15]

Out of such interactions may nature in time evolve the capability for prokaryote to eukaryote viral transfer?

In the broadest sense we are here, through the creation of wholly new gene combinations, intervening profoundly in the evolutionary process. A plausible estimate suggests that research laboratories in the United States alone will produce some $10^{15}$ to $10^{16}$ recombinant organisms per year. Industrial production could easily exceed this by several orders of magnitude, albeit probably of more limited varieties. It is unreasonable to believe that a great many of these cells will not escape our containment provisions. Such novel strains may then in a unique development broaden the base for future planetary evolution.

Can we predict the consequences? Except in the most general terms we are ignorant of the broad principles of evolution, of the factors that govern its rate and directions. We have no general theorems to account for the spectrum of organisms that we see and the gaps in between. In the microbial world, for one particular instance, what is the advantage of the botulinus toxin to the botulinus organism? Related strains seem to do well without it.[16] Why is there no coliform that has this toxin? Did evolution simply never happen upon this path? Or was it always so lethal as to prevent the development of a successful host parasite relation? We simply do not know.

We are ignorant of the relative importance of the various factors we currently perceive to participate in the evolutionary process. Major controversies swirl about the relative importance of neutral or advantageous mutations[17]--of mutations of structural genes or mutations of control elements[18]--and over factors that lead to conservation of gene order or that facilitate gene rearrangement.

We are ignorant of any absolute measure of adaptation. We are ignorant of the depth of security of our own environmental niche. How many microbes or viruses now exist that are one mutation away from human pathogenicity or two or five or one gene or two? We do not know.

In this new domain into which we leap we are surrounded by terra incognita. Areas of investigation that formerly seemed of little interest are now seen from this new perspective to be of major importance. And of course the new techniques provide powerful means to explore these areas. But while we reconnoiter, is not great caution advisable?

I know that some do not believe these organisms we now invent are truly novel. They postulate that nature has experimented with the potential of eukaryotic DNA in microorganisms for a long time. This just might be true, although the evidence is very limited. There are some curious instances that may profitably be reexamined in this light. Thus the Livingstons have described an unusual microorganism that is said to produce a protein with some aspects of chorionic gonadotrophic activity.[19]

Indeed, the discovery of a class of serine proteinases in bacteria with major structural homologies to the trypsinlike mammalian proteins led Bryan Hartley to write in 1970:

> This bacterium seems to have a mammalian gene. To finish upon a note of high and not very serious speculation I would like to suggest that an ancestor of *B. sorangium* might have acquired such

a gene from an ancestor of a cow, perhaps by the accidental introduction into a mammalian cell of a lysogenic phage directed toward *B. sorangium*. Such a phage could then have lysogenized with the mammalian DNA and thereby picked up a serine proteinase gene. Returned to the soil by the usual route the phage would inject the mammalian gene into *B. sorangium*. In other words, the bacterium might have been infected by a cow.[20]

Obviously, we are ignorant as to whether Hartley's speculation has validity or whether this is simply an instance of convergent evolution. However, if such genetic intercourse has taken place in the past we have no idea at what rates or in what circumstances, and thus we do not know how past consequences could be compared with what we are now about.

Furthermore, one should point out that we can now create combinations of DNAs from diverse organisms such as could hardly ever, plausibly, have occurred in any natural setting.

I know that some believe we will be protected from the consequences of our ignorance by the blanket theory and workings of natural selection, which in their view will stifle all of our inventions. They assume, in effect, that in each case nature has already achieved the highest possible level of adaptation. I have little doubt that had they been aware of it, the buffalo and the dinosaur would have felt protected by the same principle. I see no reason for such sanguine belief. I would add that even if nature has indeed tried out all forms and achieved near perfectly equilibrated adaptation, that does not mean we might not introduce deeply perturbing transients.[21]

Which leads me to the last unknown to add to the list. Simply, we are in the end, ignorant of the extent of our ignorance.

This, then, is the substratum upon which the NIH guidelines rest. It is crisscrossed by the faults of ignorance, the discontinuities and lacunae of our knowledge. Any one of these might fail us at any time. Research upon novel self-perpetuating organisms is as different from prior science as was the first self-perpetuating cell from all prior abiotic chemistry. There are other dimensions of hazard here. Let me refer briefly to the second class of uncertainty--human uncertainty.

Knowledge is power. As the result of the extraordinary advances in our science, biologists have become, without wanting it, the custodians of great and terrible power. It is idle to pretend otherwise. The founders of this Republic understood the dangers of power, the eternal need for the restraint of power, and they embodied their understanding in the checks and balances of our government. When they drafted our Constitution, they did not have in mind a nation of saints and angels who would need no laws, but rather a nation of fallible, temptable, even corruptible human beings, and they devised accordingly. Technology is a source of power just as much as political office. As we devise our future technologies we should profit from their example. Let us not design a technology fit only for irrational, far-sighted, unerring, incorruptible people. We must come to accept the responsibility and restraint that must accompany the power we have fashioned, or else we really may see our world dissolve in anarchy and our science with it.

There are equally important uncertainties of the third kind--moral uncertainties associated with the novel questions we now confront. We are becoming creators, makers of new forms of life, creations that we cannot undo, that will live on long after us, that will evolve according to their own destiny. What are the responsibilities of creators for our creations and for all the living world into which we bring our inventions?

A recital of risks and unknowns is lugubrious. But every risk is also a challenge and every unknown a potential for adventure. I only caution that there is a fragile line, vague and ill-marked, but fatefully real, between self-confidence and what the Greeks knew as hubris. When we are concerned with the fate not just of an individual but potentially of much of humanity, if not indeed the very biosphere, it is the course of wisdom to keep that line in full view and respect.

Let me close with a question that is the same question I had to consider a year ago. With this degree of uncertainty, with dubious footing on every side, what advice should one give to the Director of the National Institutes of Health, whose first priority is to safeguard and advance the health of our people and all people?

In my view, then, we should take every possible precaution to keep these creations out of our biosphere while at the same time seeking to obtain the scientific benefits perceived by so many as outlined by Paul Berg.

REFERENCES

1. Smith, H. W. 1975. Nature 255:500-502.
2. Anderson, E. S. 1975. Nature 255:502-504.
3. Curtiss, R. III. 1976. Annu. Rev. Microbiol. 30:507-533.
4. Bryant, M. P. 1972. Am. J. Clin. Nutr. 25:1485-1487.
5. Smith, H. W. 1976. Ciba Foundation Symposium 42, "Acute Diarrhea in Childhood," pp. 45-64.
6. Harries, J. T. 1976. Ciba Foundation Symposium 42, "Acute Diarrhea in Childhood," pp. 3-16.
7. Formal, S. B., P. Gemski, Jr., R. A. Giannella, and A. Takeuchi. 1976. Ciba Foundation Symposium 42, "Acute Diarrhea in Childhood," pp. 27-43.
8. Gray, E. A. 1975. J. Appl. Bacteriol. 39:47-54.
9. Shichiri, M., A. Okada, R. Kawamori, N. Etani, Y. Shimizu, M. Hoski, Y. Shigeta, and H. Abe. 1973. Endocrinology 93:1371-1377.
10. Boroff, D. A., and B. R. Dasgupta. 1971. Chapter 1, pp. 1-68. In S. J. Ajl, S. Kadis, and T. C. Moutie (eds.), Microbial Toxins, vol. IIA. Academic Press, New York.
11. Cohn, W., and E. Volkin (eds.). 1976. Progress in Nucleic Acid Research and Molecular Biology, vol. 19. Academic Press, New York.
12. Chamberlin, M. 1974. Annu. Rev. Biochem. 43:721-775.

13. Chambon, P. 1975. Annu. Rev. Biochem. 44:613-638.
14. Perry, R. P. 1976. Annu. Rev. Biochem. 45:605-629.
15. Huberman, J. A. 1975. Annu. Rev. Genet. 9:245-284.
16. Oguma, K. 1976. J. Gen. Microbiol. 92:67-75.
17. Crow, J. 1972. J. Hered. 63:306-315.
18. Wilson, A. C. 1976. Pages 225-234. *In* F. J. Ayala (ed.),
    *Molecular Evolution*. Sinauer Associates.
19. Livingston, V. W., and A. M. Livingston. 1974. Trans. N.Y.
    Acad. Sci. 36(2):569-582.
20. Hartley, B. S. 1970. Philos. Trans. R. Soc. London B257:77-87.
21. Fenner, F. J., and D. O. White. 1976. *Medical Virology*, 2d ed.
    Academic Press, New York. (Especially Chapters 10 and 11.)

# DISCUSSION

GEORGE WALD, Professor of Biology, Harvard University: I have for
Paul Berg, first of all, a small question and a big one. The small
question is significant, I think. I would like to know Paul Berg's
definition of a molecule--not for nitpicking reasons, but because
I think his use of this word is covering a real weakness. I would
like to know how he can possibly call the whole genome of *E. coli*
a molecule with, as I understand he said, a million subunits, each
of which is a gene. So, that is my small question.

My big question involves his apparent assumption that if he has
a gene map of the human genome he can proceed to cure diseases. We
happen to have known by now for 25 or 30 years what the defect is
in sickle-cell anemia: an exchange of glutamic acid for a valine.
I would like him to just give us a short scenario for how to proceed
after he has his map to a sickle-cell anemia or any other such
situation.

BERG: First of all, I think it is a semantic problem to define a molecule. In my view the *E. coli* chromosome, a plasmid or a phage
chromosome are molecules of DNA. Each can be isolated as a single
molecular entity. Their physical properties can be readily studied.
Their chemical structures can be defined, and at one level, their
physical structures as well; therefore, I would call these molecules.
The bits and pieces derived from them can also be defined as molecules.
These too have definite physical structures and nucleotide sequences
and can be isolated as discrete entities. Their subunits, amino
acids and nucleotides, can also be defined as molecules. I see no
inconsistency or difficulty conceptually in dealing with a complex
structure as a molecule. In short, a gene is a molecule!

WALD:  TMV is a molecule, too.  You see, Paul, we are going to need more words if this is your use of molecule.  But what bothers me is telling the general public, you see, we isolate a molecule, and we put just one molecule into *E. coli*.

BERG:  I am not trying to deceive anybody, George.

WALD:  I understand that, but your use of the word *molecule* is, you should please realize, a sort of new invention for this purpose. When you say you have something in which all components are identical, that is something you not only don't know, but there is great evidence against it.  Falkow tells us that there are innumerable antigenic strains of *E. coli* to be isolated from the human gut.  The molecules you are talking about have neither one molecular weight, one constitution--

BERG:  That is not true.

WALD:  You just said that they do.

BERG:  No, no, I am sorry, George.  I think you don't understand it. If you isolate a segment of DNA cloned in the way I described, every plasmid molecule isolated from the clone of bacteria will have an identical sequence of DNA.  You can resect that segment, and you can show that they are a homogeneous population of defined length and sequence.  That is a molecule, and every one of the cells in that clone will have that kind of molecule.  Now if you take a mixture of bacteria, each of which has picked up a different segment, then you have a heterogeneous population of molecules; but each clone, each colony of bacteria derived from a single cell will carry only the molecules which were introduced into that original single cell.  That is the power of the technique.  You can isolate pure populations of molecules containing genes in virtually unlimited quantities.

DONALD BROWN, Director, Carnegie Institution of Washington:  I would like to ask Dr. Sinsheimer a question after his exceedingly negative and depressing view of the future of this research.  It struck me that there were very few possible experiments that scientists might do to dispel some of the terribly anxious ideas which you brought forth, and since you are a really very distinguished scientist yourself and have done research in DNA, I wonder if you yourself have thought of any particular experiments which might make you feel a bit more comfortable with this research?

WALD:  Mr. Chairman, Paul Berg dealt with my little question but not my big one.

BERG: I would be presumptuous, George, if I thought I could cite all the possibilities of how to apply the knowledge we can attain by the procedures I described to the curing of all genetic diseases. But let me address myself to the one example that you gave.

We know from studies with the *E. coli* chromosome that understanding of the mechanisms of regulation permits us to determine and then define the conditions for differential expression. We can turn genes off under certain conditions and turn different sets of genes on under other conditions. By understanding the regulatory mechanisms it may be possible in some disorders to turn off defective genes and turn on alternative genes that can provide the mission function. These alternative genes, ordinarily, may be ones that are not expressed except during embryonic life. As an example of that approach Dr. Nathans suggested that if the gene for fetal hemoglobin could be turned on and the production of the defective sickle turned off some sickle-cell patients might survive and lead a normal life. This has nothing to do with genetic manipulation, that is, of introducing new genes. To understand the basic mechanisms of gene regulation we must first understand the structure and organization of the chromosome; from that information I am confident we will better understand the nature of disease and thereby be better able to undertake a rational treatment. I don't profess to know now all of the ways the information will be applied but I have the unshakable conviction that from the basic knowledge will come important and momentous benefits for medicine.

Thirty years ago we could not have answered many questions about *E. coli*; now it is the best-understood living organism on our planet. That came from studies that can now be done on animal and human cells.

WALD: You are asking both very general questions and very particular ones. I wonder if one could answer these questions in the case of *Drosophila* or, to come still closer, frogs, whether one has not already learned the turning on and turning off mechanisms, for example, that are probably applicable to the human situation.

BERG: But I just showed you all of the surprises which have come up in just the last year or two indicating novel kinds of genetic organizations and arrangements and, therefore, indicate novel regulatory mechanisms. Yes we can learn from *Drosophila*, and we can learn from the sea urchin. But already we can see unexpected diversity in those two species and I believe we can expect that there will also be unusual features characteristic of the higher mammalian organism. Unless we study such systems there is no way we are going to acquire that information. You proposed last night that we continue to synthesize *E. coli* genes and put them back into *E. coli*. But that will tell us absolutely nothing about the eukaryote chromosome.

WALD: I think the central problem before us in this direction is how many normal, healthy persons to put at risk in order to achieve the possibility, not at all clear, of eventually proceeding to cures.

BERG: I think that is one of the questions we are addressing.

BROWN: Perhaps I ought to repeat the question that I asked of Dr. Sinsheimer a few moments ago. It seemed to me in your description, your litany of all the terrible scenarios which are possible-- they are so exceedingly abstract that it is rather difficult to define experiments which might elucidate some of the information that would make you feel more comfortable, and I know you may have thought about this--I would really like to hear some of the information that you would like to see uncovered that would make you feel more comfortable with this research.

SINSHEIMER: I was not trying to be depressing; I was trying to point out the areas of our ignorance which it seems to me surround our present position. But you put your finger on an important point: there are so many possible scenarios with unfortunate consequences that it is, to my mind, very difficult, if not impossible, to devise a small number of experiments that would enable one to be fully at ease.

This particular technique does seem to me to put us in a domain which makes it very hard to rule out a whole set of possibilities. You are putting some new genes into a cell which to begin with has several thousand genes, several thousand proteins, and I don't know how many hundreds of metabolites present in a balanced set for that organism. But now you introduce new genetic components, and how you can ever predict what the resultant interactions will be, it seems to me, is very difficult. I would feel more sanguine about experiments where you are putting in one precisely and very-well-defined genetic component--let us say specifically the gene which you have previously characterized as being specifically and exclusively the gene for hemoglobin--than putting in some random unknown set. I think that is a different order of magnitude of difficulty. I think that is probably a very unsatisfactory answer to you, but I don't have a better one.

JUDITH RANDAL, Science Correspondent, *New York Daily News*: First of all, I would like to suggest to the speakers that there are people in this audience who do not understand the terms *eukaryote* and *prokaryote*, and please to explain.

It has been suggested by people on all sides of the issue that these are events which may occur in nature. Assuming this to be correct, for the time being, and assuming that this causes little disturbance as they occur in nature, I wonder if both speakers would address the issue of what happens when you scale up by manyfold, by man's introduction of many billions of bacteria. We learned from the swine flu epidemic that when you immunized 5,000 volunteers with

84

a vaccine, you got zero cases of Guillain-Barré syndrome; we now know that you would need to inject over 100,000 to have even a 50 percent chance of getting one case of Guillain-Barré syndrome. It was not until we had immunized 42 million people that we got sufficiently significant numbers to know what the outcome was. It seems to me that there may be some philosophical analogue here, and I wondered if both of our speakers would discuss this.

SINSHEIMER: I think it is an important point. There are thresholds. There are nonlinearities in nature. There are variabilities. While something may happen in nature at some degree of variability with no or minor consequence, increasing the probability of it happening by orders of magnitude may produce quite a different result. That is, if a particular recombinant organism is produced in nature at some unknown frequency, one has to ask what is the probability that that organism produced at that time will find an ecological niche in which it can multiply and develop. It may be very small, in which case it will die. If you keep repeating the experiment or if you do it 1,000 times or 100,000 times, at some point it may find that niche, and thereafter spread and evolve. This has presumably happened many, many times in the history of evolution. It is a standard consideration in the quantitative theories of evolution, and it is basically a threshold-type phenomenon.

BERG: I can make one comment. It is an extremely difficult problem to try to answer an infinite number of possibilities. Dr. Sinsheimer has said, "What if," and you can go on with "what ifs" eternally, and there is no way to answer all possible "what ifs." Some experiments are being done. One in particular has been carried out by Ronald Davis and illustrates the kind of data that led to certain of the conclusions in the guidelines.

This experiment involves a very complex mixture of recombinant lambda bacteriophages carrying a very large assortment of different DNA segments from yeast. This mixture of recombinant phages was propagated as a mixture and maintained in the laboratory. But the introduction of a single parental phage particle into that population led to the complete elimination of the recombinant organism within a few generations. Only the parental type survived. That experiment illustrates that the parental phage's chromosome outgrew and eventually eliminated the recombinant phages from the population. In short, the parent natural organism was more fit than any of the constructed phages.

To extrapolate to all possible recombinants would be unreasonable, but it is certainly an experiment that could be repeated with many different types of DNA inserts each. Although it could never eliminate all of the possible objections Dr. Sinsheimer has raised, we could, in time, be more and more reassured of its safety.

STANLEY COHEN, Professor of Medicine, Stanford University:  As many of
you know, Stanford University and the University of California have
applied for a patent on some of the techniques that are the subject
of this Forum.  Although university policy would ordinarily enable me
to share in patent royalties, I have voluntarily given up all rights
that I might otherwise have to royalties that might result from such
a patent.  I also have an outside consultantship with the Cetus Corporation, and I might add that my views on this issue long preceded my
consultantship.

I would like to now comment on some of the disturbing points that
Professor Sinsheimer has raised.  In all fairness, it should be noted
that his statements concerning our ignorance about possible long-range
outcomes of this research can be similarly applied to virtually any
area of scientific endeavor.  We can state just as well that we are
largely ignorant or entirely ignorant of the long-term effects of
the use of vaccines on viral evolution and we cannot predict what
new virulent viruses might be promoted by the use of vaccines.

We are ignorant of the effects of the construction of hybrid
plants on the plant ecology of the world in the long term.  Do we
know the long-range effects of these new plants in altering evolution and in diminishing the propagation of natural plants?

We are ignorant of the effects of the use of antibiotics on the
ultimate ecology of bacterial flora.  We are even ignorant of the
long-range effects of chemical mutagenesis of bacteria which Dr.
Sinsheimer, Dr. King, and a number of others in this room have used
in their own work.  Who can say with certainty that by mutagenesis of
*E. coli* viruses in their experiments, there is no possibility of
altering the host range of the virus so that it will then be able
to infect *Clostridium botulinum*, pick up a toxin gene, and then bring
it back into *E. coli*?  These are things about which we are ignorant.
Obviously, we may try to find out the answers to scientific questions,
but just to sit and lament our ignorance seems to me to be terribly
nonconstructive.

Now, Dr. Sinsheimer has proposed the use of safeguards and I would
agree.  Everyone agrees that this research, as well as any other research where there are unknowns, should proceed in a most judicious
and cautious way, but again I would like to repeat Dr. Brown's
question:  What sort of evidence would Dr. Sinsheimer accept as being
reasonable evidence to indicate that perhaps the hazards that he is
concerned about are not as great as he would imply?

One can say that in the past four years recombinant DNA molecules
containing genes from a wide variety of species have been introduced
into *E. coli* and multiple billions of such *E. coli* have been grown
without known hazard.  Of course, one can still argue that a combination that seems safe today may prove hazardous tomorrow or that the
next molecule that one constructs from the same species combination
may be hazardous, but these kinds of unknowns extend to all kinds of

scientific research.  The nature of scientific research implies the
inability to predict in advance completely what the outcome of the
experiment will be.  This inability does not just extend to recombinant
DNA research, but to all biological experimentation with reproducing
organisms.  If one could predict the outcome with certainty, there
would be no point in doing the experiment because you would know
the answer.  I guess what I am saying is that there will always be
questions that one cannot answer with certainty.  But I would like
to ask Dr. Sinsheimer what sort of evidence he would accept to make
him feel more comfortable about these experiments.  Since he has proposed that they continue, presumably he believes in the merit of these
experiments and in their scientific value and their benefits to society,
or at least he has been quoted as saying this.
     What sort of data would relieve some of his anxieties?

SINSHEIMER:  I don't think you really want to relieve my anxieties, Stan.
     Anyway, you are correct.  I do accept the thought that there are
positive things to be learned by these experiments.  I do not want
to see them banned.  The information to be obtained is of extreme
value.  I think you confuse the issue a little bit when you say that
when you do an experiment you don't know how it is going to come out.
That is why you do it, of course, but the concern here is that there
may be gratuitous and unfortunate side effects which have nothing to
do with the fact that you don't know how the experiment is going to
come out in the first place.
     Furthermore, to compare this with the use of antibiotics or vaccines,
I think, is a little unfortunate, too, because there you are trying
to cope with an immediate medical problem, and you are willing presumably to take certain risks in order to cope with the problem at
hand.
     To compare this with the long-term effects, for example, of hy-
brid plants, again, you are correct.  I think there could be long-term
evolutionary consequences of the introduction of agriculture and crop
systems over most of the planet.  But again you have an immediate
problem; you have got to feed the existing population.
     Bringing up some points about chemical mutagenesis--and maybe some
of my own sins in this regard--merely points out, in my own view,
that we have done a lot of things in this field and in other fields
of biology that probably were not too smart; we probably should take
more precautions about some of the kinds of things that we do and
do them more carefully and with more aforethought.
     I cannot give you a simple answer as to what kind of experiment
would really put me at ease with these particular techniques because
the concern is not just that something may happen tomorrow or next
week but that we may find out ten, or twenty, or fifty years from
now that we did something very unfortunate and irreversible.  That
is why quite honestly I would feel that this work should be done,
as I have indicated before, under maximal containment facilities,

and that should be the way in which it is done, at least in the in-definite future.

RUTH HUBBARD, Professor of Biology, Harvard University: I want to raise an issue which is implicit in a lot of this discussion. In a sense Paul Berg's presentation was a celebration of the kind of reduction-ism that molecular biology is about and that a number of people argue with: if you know more and more and more about smaller and smaller and smaller units, this will increase our knowledge about how to cure disease and be of great general benefit to humanity. It may, indeed, teach some of us who are curious about these things more, but the linkage between that and the benefit of humanity is a very, very loose one. There is a particularly bad aspect of this which all of us who are in the business of writing grants have gotten used to: at this point we say, "And it will cure disease." I really think that the implication that the curing of cancer, of diabetes--"we have now cured thalassemia, we will tomorrow cure world hunger"--by this kind of accumulation of detailed knowledge is grossly irrespon-sible.

A specific comment on the area in which the uncertainties incur risks. Paul Berg showed us slides of mammalian chromosomes and *Drosophila* chromosomes and pointed out the huge distances, the huge gaps there are between the known genes. That is exactly the problem. When he clones a gene, what he is cloning is a piece of DNA that con-tains a gene that he has characterized. He does not know, and no one else knows, what else is contained until that information is ex-pressed. It is the expression of that unknown information that some of us worry about when we raise questions about the safety of this kind of research.

BERG: First of all, I reject your argument that it is grossly irrespon-sible to point out that the kind of analyses that I was describing would lead to major benefits to mankind. It just happens to be the difference between your view and my view, Ruth. I respect the direction your research takes on the things that you feel are impor-tant, but I think it is irresponsible on your part to label other approaches as being irresponsible. They are just different than yours.

What I was trying to point out is that by careful study of such segments one can, in fact, define the nature of the genes on small segments of the DNA. Nobody is talking about continuing or blindly going ahead dealing with extensive segments of DNA of unknown genetic constitution. In fact, if you read the guidelines carefully you will see that until one knows and can define the precise genetic functions of segments of foreign mammalian DNA, one cannot remove them from the highest level of containment. In the case, for example, of the *Drosophila* segment that has the histone genes, one can say with assurance exactly what genetic information is on that segment and what the segments that are not coding for proteins must be doing as one defines lengths, number of nucleotide units, and so on.

I don't think anybody is proposing to allow people to clone large segments of human DNA, and to work with them in open laboratories. Those experiments are now essentially forbidden. Principally for the reasons of the uncertainty about what could be on such segments of human DNA, such experiments can only be done in facilities that are not yet available.

HUBBARD: There were experiments with nonhuman organisms, to wit, *Drosophila*, which are permitted, and we are not the only animal on this planet. We can louse up the environment, as well as our own health. As for your first comment, it is not irresponsible to design experiments and to do them; but to claim for them curative benefits that are extremely farfetched, I do believe is irresponsible.

BERG: I don't think they are farfetched, and therefore I reject your charge that it is irresponsible.

LUTHER S. WILLIAMS, Department of Biological Sciences, Purdue University: Taken as a minimum condition, it occurs to me that assuming the experiments were conducted with reasonable or hopefully effective containment, would the question with respect to the risks and benefits of recombinant DNA experiments be aided by deliberate and methodical experiments to answer as well as possible with the experimental data the questions posed? For example, what are the problems with *E. coli* as a host? Are there very serious difficulties derived from the fact that *E. coli* can have genetic transfer with perhaps thirty or forty other strains?

SINSHEIMER: The question, it seems to me, is similar to the one which Dr. Brown asked. Are there experiments which could be done which would resolve some of the concerns? Certainly partially I would imagine so. The difficulty is that in a sense, as has been pointed out, every time you introduce some new set of ten or twenty unknown genes into *E. coli* you don't know quite what the properties of the resultant organism are going to be. In a sense I almost sometimes feel, and I will say this partially facetiously, but one should not continue to call such organisms *Escherichia coli*. Maybe you should call them *Excelsior coli* or something like that, but you really have changed the organism.

BERG: I would like to comment. I think there has been a common confusion attributing to *E. coli* K-12 all of the properties associated with the species *E. coli*. There are experiments that demonstrate important differences between *E. coli* K-12 and normal *E. coli*. Certain plasmids carry genes which are lethal to chicken, calf, pig, etc., populations. Infection of such populations with *E. coli* carrying these plasmids has devastating consequences generally, killing the animals with severe intestinal disorders.  But the introduction of the identical plasmid into *E. coli* K-12 and infection of the same animals with these organisms is innocuous.  Dr. W. Smith, in England, who reported these experiments, is quite confident that *E. coli* K-12 cannot survive in the intestinal tracts of the animals he has studied. They obviously do not colonize the gut nor do they transfer the plasmid to intestinal *E. coli* because if they had the lethal genes would have been expressed.

Now, does that allow us to extrapolate such a result to all possible segments that could be included in such plasmids and to all possible individuals on this earth.  Certainly more experiments are needed and these will be carried out in time.  I believe that they can answer the questions that Dr. Sinsheimer has raised, but it is irrational and destructive to demand that all the questions posed by Dr. Sinsheimer's fertile mind must be known before one can move ahead at all.

# THE STABILITY OF BIOLOGICAL SPECIES

Francisco J. Ayala

*Professor of Genetics, University of California, Davis*

## THE PROCESSES OF EVOLUTIONARY CHANGE

Evolution consists of changes in the genetic constitution of populations. The raw materials of evolution are alternative genetic variants provided by the processes of mutation and recombination. Genetic variants increase or decrease in frequency from generation to generation as a consequence of two processes--random genetic drift and natural selection.

Mutation and recombination are largely chance processes. It is possible to determine the frequency with which a certain event occurs, but not when or in what individual it will occur. Mutation and recombination are accidental or chance processes in still another sense most important for evolution--namely, they are not adaptively oriented, they occur independently of whether they are beneficial or harmful to their carriers.

Genetic drift is simply the process of sampling variation from one generation to another. The smaller a population is, the greater the effects of random drift; but over many generations, genetic drift may have considerable consequences even in large populations. Genetic drift is a truly stochastic or chance process, and like mutation and recombination, it proceeds independently of the needs of the organisms.

Mutation, recombination, and drift are chance processes, but evolution on the whole is not, since it results in organisms that are highly organized systems finely adapted to the environments in which they exist. The directional process in evolution is natural selection, which promotes changes in populations in the direction of increased adaptation to the environments where the organisms live.  What genetic variants happen to be present in a species determines the boundaries within which natural selection may occur; but in what direction a species will evolve out of multiple possible alternatives is determined by natural selection.

Natural selection may be simply defined as differential reproduction. That is, natural selection takes place whenever organisms carrying a certain genetic variant have a greater probability of producing progeny than the carriers of alternative genetic variants.  The favored genetic variant will, as a consequence, increase in frequency from generation to generation.  The probability of leaving greater or lesser numbers of progeny is determined by the effects that a genetic variant has on its carriers:  the probability is increased whenever those effects are useful as adaptations to the environments.  It is thus that, overall, natural selection and evolution have produced highly organized and well-adapted organisms.

The point to be emphasized is simply this:  although the alternative ways in which evolution may proceed are delimited by the variants actually present in a population, which direction is actually followed depends on the process of natural selection.

The definition of natural selection that I have given as "differential reproduction" is deceptively simple.  Which genetic alternatives are favored in a particular instance depends on complex interactions between the genes of a given organism, between the organisms of the same species with each other, and with organisms of other species, and between organisms and the environment.  Therefore, the specific outcome of natural selection is often difficult to anticipate, since the possible outcomes are multiple.  Natural selection may favor change in one case, but the status quo in another; it may increase or may decrease the levels of genetic variation; it may favor the intermingling of different populations or their divergence into different species.  The difficulty in anticipating the outcome of natural selection is a matter of relevance to the considerations that follow.

To summarize:  The change processes of mutation and recombination have considerable import in evolution because they provide the raw materials for natural selection.  Any other natural or artificial process that injects new genetic alternatives in natural populations may also be of evolutionary consequence because it may provide additional candidates for evolution.  Nevertheless, the direction of evolution is determined by natural selection, which proceeds toward increasing the adaptation of the organisms to the environments in which they live.


THE COADAPTATION OF GENE POOLS

The previous general considerations are, I belive, appropriate before facing the question at hand:  What are the possible or likely evolutionary consequences of man-produced recombinant DNA molecules? I will focus my remarks primarily on the evolution of eukaryotic organisms, particularly those with sexual reproduction such as the mammals.

The artificial insertion of DNA molecules into the genome of an organism may bring new alternative variants in populations, and may also augment the size of the genome (i.e., the amount of DNA per cell). These effects, however, come about by natural processes as well. Mutation and intragenic recombination produce new genetic variants; augmentation of the genome takes place by unequal crossing-over. These are omnipresent natural processes, but an important difference between the natural and the artificial processes must be made clear at the outset. Artificial DNA molecules may provide organisms with sequences that would have never come about by natural processes. Genetic recombination only produces simple linear rearrangements of the DNA sequences already present in a given organism; mutation may replace, add, or delete one or a few nucleotides, but the probability that, say, the gene coding for cytochrome *c* in man would change through mutation or recombination into a gene coding for a bacterial cytochrome, or for a human hemoglobin, is effectively nil.

Before proceeding any further, it must be noted that in multicellular organisms inserted DNA molecules will be of no evolutionary consequence whatsoever unless they are inserted in the germ line, i.e., in the sex cells or their precursors. DNA insertions into the somatic but not the gametic cells will not be passed on to the following generation of the organisms, and thus will not affect their evolution.

I will not evaluate here the technical feasibility of inserting recombinant DNA molecules in the germ line of higher organisms. (Nor will I consider the possibility that the process might take place naturally, because this would not substantially affect the argument.) Let us assume that recombinant DNA molecules can be successfully inserted in the germ line of higher organisms. Is it likely that such foreign DNA molecules will be permanently incorporated into the genome of a eukaryotic species, and thus modify the evolution of the species? I believe that this event is *on the whole* extremely unlikely, although it might occur in certain limited circumstances. Let me expand, justify, and qualify this statement.

In higher organisms, the "gene pool" of a species (i.e., the sum total of the genes that make up the species) is a highly coadapted system. The genetic variants that are preserved or enhanced by natural selection are those that interact well with all other genetic variants present in the species; any genetic variant that is disharmonious is selected against and rapidly eliminated.

The evidence for the coadaptation of gene pools is pervasive (see, e.g., T. Dobzhansky, *Genetics of the Evolutionary Process*, Columbia University Press, New York, 1970). The coadaptation occurs to a certain extent within local populations. Experiments with *Drosophila* and other organisms have shown that chromosomes obtained through laboratory recombination between chromosomes from different geographic origin often reduce the fitness of their carriers, and thus are selected against by natural selection. Recombination between chromosomes from the same local population, on the other hand, results in well-adapted chromosomes. This occurs precisely because the genetic contents along chromosomes from the same population are coadapted; those along chromosomes from different populations are not.


REPRODUCTIVE ISOLATION BETWEEN SPECIES

The lack of coadaptation between the gene pools of different populations of the same species is relatively trivial compared to the lack of coadaptation between the gene pools of different species, which is quite substantial as evidenced by the universal occurrence of reproductive isolation between species. Species are reproductively isolated populations. Reproductive isolation is maintained by "reproductive isolating mechanisms" (RIMs), such as sexual isolation, diversity of ecological preferences, hybrid inviability, hybrid sterility, and others. In general, RIMs operate so as to avoid crossings between individuals of different species, and to eliminate the products of such crossings whenever these take place at all.

The formation of new species requires as a rule that populations be first geographically isolated from each other, so that no gene exchange takes place between them. As geographically separated populations become adapted to the local conditions, they become genetically differentiated. If this differentiation proceeds far enough, the gene pools of the populations will no longer be coadapted with each other. If the populations come again into geographic contact, any hybrids produced will be inviable, or sterile, or have reduced viability or sterility--that is, the populations will exhibit incipient RIMs between them. Natural selection will in such cases directly favor the development of additional RIMs, particularly those called "premating" RIMs (such as sexual isolation and ecological diversification), which avoid altogether the occurrence of interspecific crosses.

Selection favors premating RIMs because whenever hybrids are ill-adapted, any genes increasing the probability of matings within the species--and decreasing the probability of matings between species--are passed on to the following generation with higher frequency than alternative genes with the opposite effect, since the latter are more often present in the ill-adapted hybrids. Natural selection favors the development of RIMs precisely when the gene pools of different populations are not genetically coadapted with each other. In brief, the genetic pools of different species are not coadapted with each other, and their internal genetic cohesiveness is protected by a variety of RIMs.


LIKELY FATE OF INSERTED DNA SEQUENCES

Because of the internal coadaptation of gene pools, genetic materials artificially introduced in a higher organism are likely to be selected against, and rapidly eliminated, whenever the inserted DNA has important biological consequences, whether it consists of structural genes, control elements, or both. Any substantive alteration of the physiology of the species is extremely unlikely to be accepted by natural selection. The organisms carrying the inserted DNA will in such cases be at a selective disadvantage relative to the rest of the species. Therefore, the inserted DNA will suddenly or gradually be eliminated.

The insertion of DNA might be tolerated if the DNA is not genetically active in the organisms. If the inserted DNA is not translated nor affects the regulatory mechanisms of the organism, it might be effectively "neutral" with respect to natural selection, and be passively carried in the genome of the species. The opportunity would then exist for the inserted DNA to evolve gradually, through the accumulation of mutations and selection, and become eventually functional. Nonfunctional DNA, however, is often introduced in the genome of sexually reproducing species through unequal crossing-over. The artificial insertion of nonfunctional DNA sequences might at most accelerate the rate at which new, nonfunctional DNA is added to the genome. This is unlikely to have major evolutionary consequences.

Up to what extent nonfunctional DNA accumulates in a species is far from known. The accumulation cannot proceed without bounds, since as the amount of nonfunctional DNA increases, it may become a burden (and therefore be selected against) because of the energy required to synthesize it. It might be worth noting that most, perhaps all, eukaryotes already contain supernumerary DNA, i.e., more DNA than can be accounted for as carrying genetic information.

An inserted DNA sequence might be genetically functional and yet favored by natural selection if it interacts well with the rest of the genome of the species--which implies that the inserted DNA be of no drastic biological consequences, but rather operate within the morphological, physiological, and behavioral framework of the species. We may consider two alternatives (out of a continuous spectrum).

The first alternative is when the inserted DNA is simply a new allelic form of a preexisting gene. While this might improve the lot of a species in some particular situation, it is unlikely to have major evolutionary consequences. It is worth pointing out that sexually reproducing species carry in their gene pools truly enormous stores of genetic variation; under most circumstances, increasing the amount of genetic variation does not appreciably change the course of evolution.

The second alternative is more interesting, namely the introduction of a gene (or very few genes) that would give the species a new useful property without destroying the integrated development of the individual or the harmony of the gene pool of the species. This might allow the species to invade new ecological niches, i.e., to exploit new resources, or to exploit its regular resources more efficiently. A conceivable example might be the acquisition of the ability to fix nitrogen by a herbaceous plant.

It is difficult to evaluate how likely it is that a new gene of this nature might be accepted by natural selection. It would be most unlikely in the case of mammals or other higher animals with complex developmental patterns. It might be less unlikely in developmentally simpler organisms, such as plants, but even here I doubt that under natural conditions the carriers of the new gene might not have the burden of a trade-off that would place them at a disadvantage relative to other organisms of the same species. With the intervention of man, however, such as in agriculture, the incorporation might be maintained, and with great benefit to mankind. I need not discuss this last alternative, since it will be the subject of a future session of this Forum.

If the inserted gene(s) would make the species successful even without the intervention of man, the most important consequences would be ecological rather than evolutionary. The new organisms could become a pest, and displace other species from their ecological niches; or they might become effective parasites that would destroy their hosts. Let me reiterate, however, that I consider most unlikely that the incorporation of a totally new property or trait could be preserved in a higher species without man's continuous intervention.


HYBRIDIZATION AND POLYPLOIDY

I have pointed out above that RIMs are barriers against the mixing of the genetic materials from different species and that such barriers develop so as to protect the internal coadaptation of each species. Two apparent counterexamples could be advanced: (1) the natural occurrence of interspecific hybridization, and (2) the large number of allopolyploid species found in certain plant groups.

Interspecific hybridization is not a common phenomenon, but it is not a very rare event in nature either. In animals, interspecific hybridization occurs only between very closely related species, usually in places where two species have recently come into geographic contact as a result of natural, or man-caused, major ecological disturbances, as in the so-called "suture zones." These instances of hybridization support, rather than contradict, the arguments advanced above. Interspecific hybridization is usually limited to a very narrow zone along the line of contact, because hybrids are, as a rule, sterile or have reduced fitness. Usually, as time goes on, natural selection reinforces the RIMs between the species, and the rate of hybridization gradually decreases. "Introgression," i.e., the incorporation of genes of one species into the gene pool of another, is a rare phenomenon in animals. Whenever it happens, it involves very closely related species, and therefore genes with very similar DNA sequences and functions. The evolutionary consequences are limited.

Interspecific hybridization and introgression are far more common in plants than in animals, no doubt because, as I indicated earlier, plants are developmentally much simpler than higher animals. But, even in plants, interspecific hybridization and introgression naturally occur only between closely related species and do not involve the incorporation of totally divergent DNA sequences, nor the acquisition of traits or properties drastically different from those already possessed by a species.

Polyploids are organisms with two or more times the number of chromosomes of the ancestral individuals. Autopolyploids result from the duplication of the genome of a given species. Allopolyploids result from the addition of the genome of two different species. Polyploidy is a relatively rare phenomenon in animals, although it occurs among earthworms, planarians, and other hermaphroditic animals, and also in groups with parthenogenetic females, such as some beetles, moths, sow bugs, shrimps, goldfish, and salamanders.

Polyploidy, especially allopolyploidy, is common in certain groups of plants, particularly among flowering plants, where about 47 percent of the species are estimated to be recent or ancient polyploids. Allopolyploidy occurs usually between closely related species. The developmental processes are not greatly disturbed, because the genomes of the two species are complete and they are similar.

The artificial production of allopolyploids is not envisioned through the new methods of genetic recombination. In any case, the techniques to produce artificial allopolyploids have existed for several decades and have even been used between not very closely related species. The evolutionary consequences have been nil. The benefits to mankind have been, as in the production of triticale--a hybrid between wheat and rye--modest but not trivial.

CONCLUSION

I will summarize my position as follows. The leading process in evolution is natural selection that acts on the genetic variability arising by mutation and recombination. The insertion of recombinant DNA molecules in higher organisms would increase the genetic variability present in the species; it may be considered equivalent to an increase in the rates of mutation and recombination. This is unlikely to have major evolutionary consequences because natural selection will eliminate or inactivate foreign DNA sequences with drastic biological effects on their carriers. With human manipulation the insertion of single genes in plant crops might conceivably be greatly beneficial to mankind; but this will not be easy to accomplish. The potential dangers of a species becoming a successful pest or parasite must be kept in mind, although this is a rather unlikely possibility.

I have not considered the likely evolutionary consequences of introducing recombinant DNA molecules into prokaryotes, because this was outside the scope of my subject and will be considered by others at this Forum. Within the framework of thought advanced here, the following brief statement might be appropriate.

Prokaryotes do not have sexuality in the full sense, although genetic exchange and recombination occur by natural processes. Thus, the arguments advanced above, in terms of the coadaptation of gene pools and the reproductive isolation between species, do not apply to prokaryotes, or apply to them only in a limited sense. Moreover, compared to eukaryotes, prokaryotes have a much reduced integration of the genome, particularly owing to lack of tissue differentiation and of complex development.  Therefore the evolutionary consequences of the insertion of DNA molecules are potentially greater in prokaryotes than in eukaryotes.  The possibility of producing strains of prokaryotes either useful or harmful to mankind and to the ecological balance of nature obviously exists.


## RESEARCH WITH RECOMBINANT DNA IN EUROPE

Sir John Kendrew

*Director General, European Molecular Biology Laboratory, Heidelberg*


You may recall some discussion in the New Testament about the problems of serving two masters.  Serving ten or more is what international organizations are all about.  It has its problems.  The difficulty in Europe, as in so many other spheres, is that the region consists of a large number of sovereign states with different science policies, different legal systems, different national interests, and different national temperaments.  There are various international organizations at different levels that offer the possibility of united actions and policies; but all of these organizations have a more or less limited geographical coverage and extremely limited authority.

I want to discuss the activities of several of these--specifically the European Community or the Common Market, the European Science Foundation, the European Molecular Biology Organization--in the field of recombinant DNA.

I think it is well known that so far the only comprehensive guidelines for carrying on research in this field that has been elaborated outside the United States have been published in England in the so-called "Williams Report."  I won't discuss the guidelines in detail.  I simply observe that they differ marginally from those of the United States in placing more emphasis on physical containment and less on biological containment.

The European Molecular Biology Organization (EMBO) is a private body funded by sixteen Western European governments.  It has established a standing advisory committee on recombinant DNA,* the object of which is to try and harmonize the guidelines that may be set up either soon or in the future in the nation states of Europe.  Two of the members of that committee, Professor Charles Weissmann and Dr. John Tooze, are here and you may like to ask them some questions about the activity of that committee.


*Membership and terms of reference of this committee follow this paper.

Now, alongside that we have a body called the European Science Foundation (ESF), which represents academies and research councils but not governments; therefore, it has no legal power. It is a quasi-official body. I think it now has representatives from eighteen Western European countries. It established its own committee on recombinant DNA and will be establishing a standing committee consisting of representatives of the national advisory bodies of EMBO and the European Research Council made up of medical research councils with the object of standardizing or harmonizing practices and guidelines.

More recently the European Economic Community (EEC), which is a body of nine countries with much stronger legal powers than the ESF--indeed, the ESF has none at all--has taken an initiative in inviting the chairmen of what have now in Europe come to be called the Genetic Manipulation Advisory Groups (GMAGs) to discuss these problems. The implication is that the Community may issue a directive that all recombinant DNA research carried out in its nine countries, including industrial research, which is a very important point, should be conducted in accordance with the Williams guidelines or some very similar guidelines. It looks as if the long-term aim of the Community here appears to be to promote legislation requiring the compulsory registration of activity in the various countries, and to assume the responsibility of making periodical revisions of the guidelines.

On the face of it the European Science Foundation and the European Economic Community initiatives require a degree of coordination that has not yet taken place. The two bodies have tried to assume rather similar responsibilities. One of them, the Science Foundation, has no legal authority but a rather wide geographical coverage. The other one, the Community, has a very limited geographical coverage, but at least in principle has a strong legal authority. You might say that this kind of confused situation is only too characteristic of the disunited states of Europe. One very much hopes that an accommodation or a coordination of this multiplicity of committees and responsibilities can be achieved, because after all scientific activity in this as in other fields is an international affair that does not or should not recognize political groupings of nations. I might recall in this connection that the countries of Eastern Europe have no part in any of the arrangements I have so far mentioned.

I would like to refer, before leaving the international scene, briefly to the activities of my own laboratory, the European Molecular Biology Laboratory, which is supported by ten governments. We are constructing a containment facility to the highest standards specified either by the NIH or the Williams guidelines. It will be ready towards the end of this year and will in part be a service facility for visiting groups from the European countries to come and carry out work at containment levels and under safety conditions that they do not have available in their home institutions. In addition, of course, we plan also to have our own research program in the field.

I might make one or two brief comments on activities in individual Western European countries. In many of them now there are national committees for genetic engineering. These include France, Germany, the Netherlands, the three Scandinavian countries, Belgium, Italy, and probably others. The pattern is different in different countries. In some the committees have governmental status and have varying degrees of authority to establish guidelines, to inspect laboratories, to authorize experiments, and so on. For example, the GMAG in the United Kingdom has most of these responsibilities and authorities. In other countries the committees are purely advisory or unofficial. It would, I think, be fair to say that with the partial exception of the United Kingdom, in most of the countries the relationship between government, the law, public health, and the scientific community has not yet finally been resolved. In many countries P3 or even P4 containment facilities are being constructed or are already operational.

I have not so far mentioned similar activities in Eastern Europe, but to my knowledge research in this field is going on in the Soviet Union and in Hungary; there is also discussion in Eastern Europe of the guidelines. I have not time to talk about these activities in the countries in more detail, and those of you who might be interested to follow this up will find a very useful article in last week's issue of *Nature* about activities in the different European countries.

I now would like to turn just for one minute to the general atmosphere surrounding recombinant DNA research. I think it would be fair to say that the climate in Europe is distinctly cooler than it would appear from outside to be in the United States. There are, of course, continual discussions in the various national institutes. There have been discussions between scientists and nonscientists, scientists and trade union representatives, and so on. Many of the same kinds of attitudes that you have heard talked about in this meeting and will hear in this meeting have been expressed in various quarters in Europe; but I am glad to say that the mood in those parts of Europe that I know about anyway is not a highly contentious one, and we all hope that the generally calm spirits in which discussions have been carried out in Europe will continue.

I would like just to say a word on a topic lying outside my brief. I want to refer very briefly to activities at the world level. In intergovernmental circles at the world level the main responsibility lies with the World Health Organization, which has set up a committee charged with considering the public health implications of recombinant DNA research in the international context.

More recently, the nongovernmental world body of scientists, the International Council of Scientific Unions, has established a "Committee on Genetic Experimentation," the acronym for which is COGENE, a committee whose terms of reference on the world level are not dissimilar from those of the EMBO standing advisory committee in the European area.

Finally, having tried simply to give a factual account of what is going on in Europe, I will close with some indications of my own attitudes about these things.

First of all, in the context not of recombinant DNA but of international understanding, I must say that in spite of the complicated committee structures with overlapping responsibilities that we have in Europe and the somewhat strained relations that from time to time occur between the committees of one body and another, I am most impressed at the way in which the recombinant DNA situation seems to be settling down in Europe with a good deal of conformity and a good deal of harmony between countries.

I only wish that we could make as much progress over matters like the dates of summer time in Europe as we do about the regulations for recombinant DNA. I think that it indicates that these international relations, cumbrous though they may seem, can in some areas be quite effective.

As to the work itself, I consider it to be of the highest importance. I think it should be prosecuted with the utmost vigor and with the utmost safety, and it is precisely for this reason that I have been very glad that in my own laboratory we are constructing facilities where the work will be encouraged and the best safety precautions will be adopted.

I, personally, am a quite unashamed reductionist. I think that ignorance is the ultimate enemy. I am glad that we have a new and power- ful weapon in the fight against ignorance. I think that faced by hypo- thetical benefits which always are accompanied by hypothetical dangers, I believe that as in all scientific endeavors in the past we should maximize the former, minimize the latter. Both these things are in our power, and I hope that this technique, which in my view is of the highest importance intellectually for the future of the human race and the future of other animal species, will be fully exploited in the safest conditions in the United States, in Europe, and throughout the world.

EMBO STANDING ADVISORY COMMITTEE ON RECOMBINANT DNA

The European Molecular Biology Organization has established the Stand- ing Advisory Committee on Recombinant DNA. The composition of the committee is as follows:

Professor C. Weissmann
(Zurich, *Chairman*)

| | |
|---|---|
| Professor E. S. Anderson (London) | Dr. K. Murray (Edinburgh) |
| Professor W. F. Bodmer (Oxford) | Professor L. Philipson (Uppsala) |
| Dr. S. Brenner (Cambridge) | Dr. J. Tooze (Heidelberg, *Secretary*) |
| Professor F. Gros (Paris) | Professor H. G. Zachau (Munich) |

The terms of reference of this committee are as follows:

1. The committee, on request, will advise governments, other organi- zations, and individual scientists about technical and scientific aspects of experiments with recombinant DNA. It will consider questions of experimental containment at various levels of potential risk.

2.  The committee will explore the possibility of instituting, or arranging for, training programs in accordance with the postulated needs, in consultation with national bodies or individual laboratories if necessary.

3.  The committee will collect copies of laws, rules, and guidelines from various countries relating to and controlling experiments with recombinant DNA.

4.  The committee will maintain a close liaison with the European Science Foundation and other international governmental and nongovernmental organizations that are concerned with the various aspects of experiments involving recombinant DNA.

The EMBO Standing Advisory Committee will have no regulatory or legislative functions, which remain the responsibility of national governments and other organizations, and will concern itself only with scientific and technical questions.  If and when governments or other organizations request advice on scientific and technical matters, a meeting of the whole committee will be convened and the committee will make a written reply under the signature of its chairman.

If and when individual scientists seek the advice of the committee, the chairman will transmit the request to an individual member of the committee who will be asked to give a reply to the correspondent.  In replying the member will state that the committee has no legislative or licensing authority and that it is the responsibility of individual scientists to ensure that their experiments conform to those national and international standards in force in the country in which they work.

The committee believes that one of its chief roles will be advising individual scientists about strains of bacteria, plasmids, and bacteriophages best suited for particular experiments.  Although it has no resources to maintain a type collection the committee believes such a collection should be established in Europe either within the framework of the EMBO or elsewhere.

## THE RESEARCH AND THE PUBLIC INTEREST

### Stephen E. Toulmin

*Committee on Social Thought and the Department of Philosophy, University of Chicago*

The current debate about recombinant DNA has had the effect of crystallizing out a whole range of questions about science and the public interest that have been, so to say, in supersaturation for some fifteen or twenty years, some of them ever since Hiroshima.

These questions have to do with the public accountability of scientists for their work; with how the legitimate interests of the public are to be represented in decisions about science at a time when after some 300 years all those promissory notes about future social welfare that scientists have been issuing ever since Francis Bacon are at last within sight of redemption; with the structure, recruitment, and reward systems of the scientific community and its institutions; with whether the community of scientists must at last take on the formal responsibilities, as well as the privileges of a statutory profession in the way in which physicians, attorneys, and even beauticians have long since been required to do, and so on.

In a word, the problem of regulating recombinant DNA research has become not merely a practical issue but also a symbolic one. That being the case, it is harder and more important to sift out and focus in on those questions that arise specifically and necessarily out of the recombinant DNA debate itself, as contrasted with other more peripheral questions that, while they are important and urgent in themselves, have nothing specially to do with the immediate matters in hand.

With this in mind, I am going to set aside a great many questions that have come up in the course of the DNA debate and that need to be dealt with carefully and seriously in the course of the wider public debate about science and the public interest. I do so with some regret because many of these questions are not getting as full and informed a discussion as they need. But given the time at my disposal I must start by focusing on the two central issues directly raised by recombinant DNA. First, in what respects are scientists accountable to the public authorities, whether the Cambridge City Council or any other, for any risks attendant on recombinant DNA research? Second, does recombinant DNA research, as at present proposed, represent an irreversible, or even a hubristic, incursion by human investigators into a region of knowledge that would better be left untapped, untouched?

To begin with the question of responsibility for risks, I think one can say quite flatly that if there is any serious possibility of recombinant DNA research leading to harm among the general public, then the right of the public authorities to intervene is beyond dispute. Of course, I realize that there is a lot of disagreement among well-qualified scientists about the scale and nature of any such possible risks. My own primary training was in physics rather than biochemistry, and I have to leave it to others to settle that matter among themselves. It is up to those who understand the issues much better than I ever could to arrive at some sort of a consensus first, and they must not be surprised if everyone else adopts a cautious and conservative attitude toward the matter in the meanwhile.

But there is one extreme view of the issue which it seems to me can be dismissed out of hand. Some of my scientific friends are so shocked by the spectacle of the Cambridge City Council placing hurdles in the way of Harvard's recombinant DNA program that they have reacted by making some quite indefensible constitutional claims. They have suggested that scientific research is not merely a "constitutionally protected activity" under the Bill of Rights--that is, that the right to do whatever scientific research one thinks fit without state interference is guaranteed by the First Amendment--but that the unimpeded exercise of that right is protected absolutely. As to that there are two fairly brief things which can be said.

First, it is not certain that the First Amendment does in fact cover the right to do whatever scientific research one thinks fit (or, indeed whether it covers the right to do scientific research at all): the matter has never come up for adjudication. I, personally, would predict that a case raising this question will probably reach the Supreme Court sometime during the next fifteen years or so, and that the Court will probably decide that freedom of speech does, at least in general terms, embrace freedom of scientific inquiry. But that is pure guesswork at the moment, and in any event the Court might well write in a lot of small print limiting the application of the phrase "whatever scientific research one thinks fit."

Second, First Amendment or no First Amendment, the possession of a right is, as every first-year law student quickly learns, not the same as the exercise of that right. If we seek to exercise our constitutionally protected rights in a manner and a situation in which there is a "clear and present danger" of public harm, the public authorities are perfectly entitled to intervene and place restraints on that exercise, as Justice Holmes reminded us. The First Amendment does not authorize us to panic a crowded theater by shouting, "Fire!" where there is no fire. Given the degree of disagreement between scientists over the scale and nature of the risks involved in a DNA case, the Cambridge City Council quite reasonably apprehended the possibility of public harm, and it would have been negligent of them not to have intervened.

Why has there ever been any doubt about this? The reason is, I believe, because in one crucial respect the recombinant DNA is, in fact, a historic first. There have been previous cases in which the actual conduct of scientific experiments posed risks to the human beings directly involved. Hence all the current concern in recent years about the ethical review of research involving human subjects.

Again, there have been previous cases in which the effects of applying the results of scientific research on a large scale subsequently posed a serious threat to public health or welfare. The whole controversy about the use and abuse of nuclear technology is an obvious example of this.

Plenty of secondary lines of scientific investigation, too, have required special constraints, for instance, those involving dangerous viruses or nerve gases or radioactive substances, and the scientists who do their research in these fields are by now perfectly accustomed to the fact that they have to place certain constraints and that indeed constraints are placed on them in their research by the public authorities. But I can think of no prior case in which the actual conduct of fundamental experiments in a basic natural science itself directly posed a threat of general public harm. Even in the case of nuclear physics and the "artificial transmutation of the elements," as it was called when I was a boy, the direct effects of the initial experiments conducted by Rutherford and his colleagues at the Cavendish back in the early 1920s (not in 1942, as we were told last night), the direct effects of these experiments were entirely localized and involved no risk at all to the general public.

Why is the recombinant DNA case unique in this respect? In all previous situations involving the production by scientists of toxic substances or agents as part of a program of strictly basic research there was no difficulty in limiting the spread of those agents or sub-stances. But the very heart of the DNA problem, it seems to me, is the suggestion that any "rogue" agents produced artificially in the course of research might have the power to multiply themselves and spread throughout the population at large, e.g., by colonizing the human gut, so distinguishing themselves from, for instance, the minute quantities of artificially radioactive material produced by Rutherford's experi-ments and the like. It is this multiplication effect that is rightly perceived by outsiders as requiring special safeguards and as justifying a deliberate, conservative approach until it is clear that such safe-guards are available.

To move now to the forbidden knowledge issue, it is evidently unwise for scientists to underestimate the power of the public imagination in this respect. The manner in which the recombinant DNA issue is being presented for public debate, both by journalists and by some scientists themselves, has been one that has quite naturally awakened echoes of, for instance, the Faust legend and so provoked the kinds of public anx-iety and suspicion that in earlier centuries confronted the alchemists--for instance, the suspicion that they were attempting to produce the *Homunculus*, that is, an artificial human embryo generated within their own alembic or retort out of lifeless raw materials. So, it is necessary to correct some of the misunderstandings on which the current suspicion rests. Certainly I know of nothing in the way of DNA research that is actually in prospect as contrasted with speculative propaganda either pro or con which could even remotely answer to the *Homunculus* specification.

During the 1920s and 1930s, as I recall, the activities of the atom splitters at the Cavendish and elsewhere aroused something of the same frisson among the general public. Splitting the atom was suspected also of tampering impermissibly with nature's mysteries. Rutherford, in response, called his own little book of popular science *The Newer Alchemy*. But it is clear by now that the whole field of atomic and nuclear physics covers a multitude both of good things and of bad things. So the ques-tion is not whether to permit or outlaw all such work but rather what sorts of research we should do in this area, with what safeguards, and how we shall control the possible misapplication of the knowledge gained as a result.

I want to take the two parts of this issue in turn. First, any claim that all recombinant DNA research in itself involves impermissible tam-pering with the natural processes of organic evolution is surely far too general and undiscriminating. In itself the research does nothing of the sort. It could do so only if its by-products escaped into the biosphere or if the knowledge gained from the initial research were misapplied.

In any event, we have, as we have already been reminded, been intervening in the natural process of organic evolution for millennia. All culture, especially the domestication of plants and animals and their selective breeding, involves tampering with evolution. So, once again the question is not whether we are to begin doing it for the first time but rather how we are to be sure in this particular case that we are doing it for good rather than for ill.

Second, how are we to be sure of that? Again, I believe we have perfectly good models available that answer that question in principle, if not in detail. The principles may remain the same, but the details are always new. The issue was settled in principle, as I see it, in prehistory, and the outcome is enshrined in the legend of Prometheus. In its own time and in its own terms, discovering ways of producing fire artificially was as daunting as discovering techniques for producing artificial radioactivity, nuclear power, or synthetic forms of DNA is today.

As people very soon came to recognize, the proper response was not to outlaw the very use of fire. Rather it was to invent the legal concept of arson. This is known as replying to a four-letter word with a five-letter word. Rather, I say it was to invent the legal concept of arson and to develop effective legal sanctions, institutional mechanisms, public sentiments, and other practical safeguards against the misuse of fire.

It is very easy to express Arcadian sentiments, but I think it would be worthwhile, especially after a January in Chicago speculating about what it would be like to live without the artificial handling of fire. So, for all I can see, there is nothing forbidden or impermissible about the knowledge we could get from recombinant DNA research, provided that we are also taking seriously the practical question of the safeguards against foreseeable harm and even against unforeseeable harm and the eventual possible misuse of that knowledge.

This, of course, brings us back to all those wider questions that I set aside at the outset about the institutions of science, about their role in the larger national life and so on. Once we move beyond questions about the substance of the public interest in the recombinant DNA issue, questions about process still remain to be dealt with. For how should the interests of the larger society be represented and how have they been represented in the discussions leading up to the preparation of the NIH safeguards, for instance? And how for that matter should the interests of the larger society be represented in the discussion of other major policy issues involving the impact of new developments in scientific technology? All in all, I think a disinterested outsider is justified in saying that Paul Berg, Donald Fredrickson, and the high command at NIH did a very respectable and conscientious job of working on those safeguards in the absence of any proper institutional setup for dealing with the societal aspects of science policy.

In the absence of any properly representative forum for taking such decisions, that is, they improvised as carefully, conscientiously, and responsibly as they knew how, but the fact remains that what they did was a piece of institutional improvisation, and in consequence, whether or not the safeguards they arrived at really did proper justice to the legitimate interests of the wider public (and they very possibly did), that justice was not and could not have been seen to be done.

After all, if Prometheus, having introduced fire, claimed the right to compose the laws against arson singlehanded, he, too, would have been under some suspicion of being self-interested.

So, I end with a plea. In our concern over the substance of the risks and the other larger issues posed by recombinant DNA, do not let us lose sight of the longer-term problems about process. Since 1945, the natural sciences have moved into a new epoch, in which their conduct and priorities have become a matter of importance not merely to themselves but also to the larger society, and as a result the scientific community is having to learn painfully but inescapably to accept a new kind of accountability, and to see its programs scrutinized by social as well as intellectual, Baconian as well as Newtonian, standards. A good deal of hard feelings and conflict of interests will be circumvented and avoided if the public representatives of the scientific community acknowledge that in this new Baconian epoch the larger society does have a legitimate interest in the conduct, as well as the outcome, of research in the natural sciences. For that will mean acknowledging also the need to collaborate in the task of devising and establishing the more responsible and representative institutional machinery we need if we are to have sufficient assurance that the legitimate interests of both science and the larger society are being given their proper weight when all those further issues begin to arise in the future, of which the problem of recombinant DNA research is only one early, though difficult and contentious, example.

## DISCUSSION

JON BECKWITH, Harvard Medical School: First of all, I would like to say that I do think people can be bought, including scientists. There is plenty of evidence to show that scientists are no more pure, in this regard, than people in other walks of life.[1-3]

Secondly, I would like to suggest that even though, throughout the history of biology, the term *man* has consistently been used to refer to human beings, this practice subtly reinforces sexist attitudes in this society. Therefore, I hope people will make an effort to use *humans* or *human beings* instead of *man*.

Finally, I would like to respond to Paul Berg's and Dan Nathans' comments concerning some of the benefits of recombinant DNA research. I want to relate these comments to the issue of reductionism in science mentioned earlier by Dr. Kendrew. Clearly, the reductionist approach in biology--studying in detail very small components of biological systems--has been an important component of the remarkable progress of molecular biology. Focusing on individual genes, their structure and function, has been particularly successful. On the other hand, there are many biological problems where taking such a narrow approach will result in missing the forest for the trees. However, when the reductionist approach is applied to society's problems, it is not only myopic, it is dangerous. In fact, examples can be cited where this kind of approach by scientists to social problems has created serious problems.[4-7]

Now, if I heard Paul Berg correctly, he stated that heart disease, cancer, and some other health problems are a consequence of genes and their regulation. Further, he went on to state that our goal was to understand the human genome and that this will have profound significance for improving human health. I believe that these statements are false and represent the influence of the reductionist approach on our attitudes to more complex problems. They ignore the complex web of factors, environmental, genetic, and others, which contribute to disease, and focus on only one of these components, the genetic one. These are not politically neutral statements since they have implications for the priorities of this society. We all have been corrupted by the exaggerated claims we make for the benefits of our research, particularly in justifying our grants. The overblown promises to society made by scientists are, in the long run, a much greater threat to the future of science than the criticisms leveled at recombinant DNA research.

Given the concerns over recombinant DNA research, given the major source of our health problems in the environment and in social and economic arrangements, what we should be discussing here is which are the real needs of the society and whether the proposals which have been made here will contribute to the solution of societal problems.

BERG: I will not respond, Jon, to your comments about which are the more important questions to be addressing here. I will respond, however, to your comments about the genetic origins of heart disease, cancer, and other ills.

I think you must know that one of the most interesting and exciting discoveries of recent years is the identification of a collection of genetic diseases referred to as hypercholesterolemias. They are the consequence of defective genes which disrupt the regulation and the handling of cholesterol. As a consequence extraordinarily high levels of cholesterol are built up, leading invariably to death of the homozygous individuals and to early or premature death of the heterozygous subjects.

In one particular type of hypercholesterolemia the defect is due to the inability to regulate the intracellular production of cholesterol because of the lack of cell receptor for the serum protein that transports cholesterol. In this instance then a major form of heart disease results from a malfunction of a particular genetic element in the human genome. Understanding that defect and the impaired regulation could have a profound influence on a major human killer!

There is also the overwhelming evidence that the genetic constitution of humans plays a role in their susceptibility to cancer, virus infections, and a whole host of diseases. In my view it is absurd to deny that.

BECKWITH: I was not denying that there was a genetic *component* to diseases. What you did was to emphasize only that, and you did not mention the environmental components. In this society, particularly in the case of cancer and heart disease, the environmental components are clearly the major ones. It is, of course, convenient to emphasize the genetic components and focus on the afflicted individual, since otherwise we would have to consider some significant changes in the way we operate in this society.

BERG: The environmental agents may well be acting through the genetic components. The more information we have about the genetic components, the more we can understand or approach the questions of the environmental hazards. Quite possibly environmental carcinogens act on endogenous viruses which are carried in the genome. We will never know that unless we can make a refined genetic analysis of the components of the human chromosome.

BECKWITH: Why not eliminate the carcinogens?

BERG: That is one way, but I think we may not be able to do that entirely. Therefore the information gained from an understanding of how carcinogens act may well prove to be more important. I don't think that one wants to bet at the moment on one course or another. I think both avenues should be explored. You are trying to tell everybody here that only the carcinogen approach is the valid one, and the others are irrelevant.

NANCY ABRAMS, Office of Technology Assessment: I was hoping to hear a talk that would explain to some extent what the public interest really was. Dr. Toulmin's talk was very charming, but all it did essentially was to tell the scientific community, "Hey, just notice that there is a public." Now, I would like to say that one of the reasons I think there are so few creative solutions to this problem coming up is that there is not enough input from nonscientists. I am a lawyer and not a scientist, and I am very disturbed by the assumption that I seem to see among scientists here that if the research can be done, automatically it should be available to industry, that the only question here really is safety.

I think, for example, it would show extremely good faith on the part of those members of the scientific community who want to go ahead with this if they would support, for example, a national rule

that no chemical or drug that had to be manufactured with DNA recombinant technology could be sold in the United States or imported into
the United States. This would allow all the research to go ahead
and remove the profit incentive. Let me give you a few examples of
some of the advantages. This would be an introduction to the discussion about guidelines. Guidelines would still have to go on because
there are possibilities of dangers in the technology. But from the
point of view of the public, there are some other advantages.

One of the things people worry about is not that scientists are
careless or that they are immoral. What they worry about is that
when there is a profit motive people cut corners. This is one of the
sources of problems with the nuclear issue, and I am sure it is going
to come up with DNA technology when people find out about it. By
creating a disincentive to use this for commercial purposes, you
would eliminate the need to rush. If, in fact, the goal of this
research is knowledge, then there is no point in rushing, and there
is no way to cover it up either.

If it is going to be more expensive to do the research with good
containment facilities, then those containment facilities could, in
fact, be bought with additional funding from whoever is giving this
funding out, whether it be NIH or whoever, but it is open.

If industries are expected to do these things carefully in high-
containment facilities, who is really going to give them the money
to build these things? It is the expense of them that is going to
cause them to cut corners, and I would like to know whether the
proponents of this technology would be willing to support that kind
of a limitation which would say absolutely nothing about the kind
of basic research that could be done.

RICH: I would like to point out that later today we will have a session
devoted to industrial applications in one particular context, and I
think we will have further elaboration of some of these points.

ABRAMS: I want to know from Dr. Berg and from some of the proponents
of this technology whether it is absolutely crucial to the development
of basic knowledge that industry have immediate availability of the
results. That is all I want to know.

RICH: The problem is we live in an open society. We publish all of
our results. None of this work is secret. It is available to people,
whoever reads the literature.

ABRAMS: That is absolutely right. However, there are some problems with
industry having proprietary information which may or may not be public,
and I think that that is something very serious to be considered.

BERG: Are you asking if the people who are engaged in this research, let
us say in the university, are opposed to this work going on in commercial or industrial laboratories?

ABRAMS: I am sorry my question was so unclear. What I meant was I feel that if scientists who want to go ahead with this research would support the concept that it not be available to industry in this indirect way, that is, industry could go ahead and do the research. They just could not sell anything they made with it.

BERG: I am sorry. I cannot see the point to that.

ABRAMS: The point is that it removes the profit incentive and says, "Look, if we are basic scientists, we are doing this for basic knowledge."

BERG: If you are proposing that industry should be permitted to go ahead and carry out this research but not to apply their knowledge to useful products, then I think you are misled. You are advocating acceptance of the risks of the experimentation but foreclosing on the potential benefits that might accrue from them. It does not seem sensible to me.

ABRAMS: This is the last thing I am going to say. I am sorry. The benefits are very problematic. There have not been technology assessments, to my knowledge, of the effects of nitrogen fixation by thousands of acres of wheat in the middle of the United States or what is going to happen with those bacteria that are eating up the oil spills, where are they going to go. They may be benefits. They may not be benefits. I think there are so many uncertainties that this would be a first step in figuring out a little more whether we could get those benefits without this technology.

STUART NEWMAN, State University of New York, Albany: I would like to point out what I consider a specious analogy between processes of natural evolution and processes of animal and plant breeding and artificial selection, on one hand, and recombinant DNA research, on the other hand, that has been put forward by proponents. What is not taken into consideration here is that the normal evolutionary process and the artificial breeding process occur on the basis of selection of preexisting variability. Even if mutagenesis is involved, generally it is one nucleotide at a time, one base at a time, which is changed. This, if it occurs naturally, if subject to all the complex control circuitry in the environment that that organism is immersed in and the particular protein which is changed, subject to the complex control and feedback circuitry that that protein in its unmodified form is subject to in the cellular environment. This is quite different from the wholesale importation of a big block of genetic material from an eukaryotic or higher organism into a prokaryotic organism. In that case you have a complex set of proteins that are not unlikely to be expressed in some circumstances, that are not subject to the normal control circuitry. There is nothing in the cell that will recognize and feed back upon the products that are formed, and there is probably nothing in the environment that when the organism is modified in that way will feed back upon and control that environment.

I would like to put this in relation to some of Professor Ayala's comments, because his main emphasis was on the possibilities of recombinant DNA getting into higher organisms, and his conclusion is that it will probably be deleterious: that organism will probably not survive, will probably not proliferate and expand.

Now, I agree with that. I think those higher organisms are us, and if anything from a recombinant molecule which has been proliferating in the environment gets back into us, I don't think we have a very good chance.

On the other hand, Professor Ayala spoke about the probability of what would happen in a much lower organism, say a bacterium which receives the result of a recombinant DNA that it would not get naturally, and he speculated briefly that this could likely have much higher consequences because the control is not as rigid and it is possible that such an organism would increase its virulence, find new ecological niches and be much more mobile.

I would like to ask a question of Professor Berg, which is, I think, a valid analogy. If you were a corporate executive in the asbestos industry thirty years ago, and somebody suggested that possibly the breathing of asbestos fibers and their incorporation into the lungs of workers might have a deleterious effect, although there is no evidence out on that effect, what would your attitude be for preventing asbestos fibers from going into the lungs of individuals, just as a hypothetical situation?

BERG: I don't think I can sit up here and respond to what is an irrelevant question, and one which I think is not very helpful. Such hypothetical questions are not very useful in dealing with the recombinant DNA issue. So that you don't think I'm trying to duck your challenge, let me say that I probably would have tried to learn a great deal about the issue and about how to test the speculations, and then I would have acted as responsibly as I could on the basis of that investigation.

AYALA: I want to comment briefly that I quite agree with the first part of the comments that were made by the last speaker. I made it explicit there is a difference between genetic variation introduced by techniques such as recombinant DNA and the naturally occurring process because sequences that would never arise by natural processes may arise by this technique.

That, incidentally, makes it only most unlikely that such sequences be accepted evolutionarily. Let me point out, however, that the situation as to naturally occurring variation is not quite as you seem to see it, of a single nucleotide substitution one at a time. When we do look at natural populations, it is not unusual to find a gene locus at which, just by techniques rather crude, as electrophoresis, we find twelve, fourteen, twenty different variants existing in all high frequencies in any one given local population. So, the occurrence of natural variation is a pervasive phenomenon. It takes very little effort to calculate that the number of different genotypes that could possibly be produced out of the naturally occurring variation in any sexually reproducing species is many, many orders of magnitude greater than the number of atoms in the universe. This is just to give you an idea of the kind of variation that exists.

Now, the last part of your comment is well taken. I understand that perhaps Dr. Davis or others could consider the matter of evolution at the level of prokaryotes. My last statement made it clear that I was talking about the possible evolutionary implications for higher organisms. The terms under which evolution of lower organisms have to be considered are somewhat different, and this is about the main thing I was saying, and I was pointing out that obviously possible dangers exist. I am sort of passing the buck, but I want to make it clear that that was not my topic, so as to frame my comments.

FRANCINE SIMRING, Friends of the Earth: Before I pose my question, I would like to say that Dr. Berg and Dr. Sinsheimer, two eminent scientists, have offered us differing points of view. Dr. Berg is anxious to proceed with the research. Dr. Sinsheimer is commendably cautious, and, if I am correct, supporting the view that the research is to be limited to several centralized high-containment facilities.

However, whether or not the research should continue at all, of course, or should proceed only with the development of an alternative to *E. coli* as a host organism has not been democratically arrived at with broad public participation, evaluation of that technology, and the development of a national policy. So, there is an important question of whether we can permit one man or a small group to make decisions for 200 million Americans whose lives will be very importantly affected by that decision.

My question is directed to Dr. Berg and the rest of the eminent scientists who signed the original letter calling for an eighteen-month international moratorium on two types of the highest-risk recombinant DNA research. The names are Paul Berg, David Baltimore, Herbert Boyer, Stanley Cohen, Ronald Davis, David Hogness, Daniel Nathans, Richard Roblin, James Watson, Sherman Weissman, and Norton Zinder, a number of whom are present with us today, and so I direct my question to Dr. Berg and all of you others here with us.

In this letter these respected scientists requested of the Director of the NIH immediate consideration to establishing an advisory committee charged with, and I read from the original letter, "overseeing an experimental program to evaluate the potential biological and ecological hazards of the above types of recombinant DNA molecules." This program of experimentation to assess the hazards was not carried out, although the terms for it appeared in the *Federal Register* in the following November:

"However, the use of this technology has various possible hazards because new types of organisms, some potentially pathogenic, can be introduced into the environment if there are no effective controls. The technology is also capable of producing microbial organisms which can be useful or harmful to agriculture and industry and thus secondarily affect human health. The goal of the Committee is to investigate the current state of knowledge and technology regarding DNA recombinants, their survival in nature, their transferability to other organisms, to recommend programs of research, to assess the possibility of the spread of specific DNA recombinants and the possible hazards to public health and the environment and to recommend guidelines on the basis of the research results. This Committee is a technical committee established to look at a specific program." The program for assessment of hazards was never carried out.

At the end of the July 1974 letter is the following sentence: "Nonetheless, our concern for the possible unfortunate consequences of indiscriminate application of these techniques motivates us to urge all scientists working in this area to join us in agreeing not to initiate experiments of type 1 and 2 above until attempts have been made to evaluate the hazards and some resolution of the outstanding questions has been achieved." Some of the outstanding questions have not been answered. The hazards have not been evaluated, but the moratorium was rescinded the following February.

My question is why was this program of the evaluation of hazards not carried out before the NIH guidelines were developed, since they were to be based on this research? Why did none of the letter signers persist in what they originally called for? Since it was mandated that the recombinant DNA advisory committee base guidelines on this assessment research, why did you all not call out for this instead of calling out for promotion of the research as against the growing demand for limitation of the research?

RICH: I will call on Dr. Singer to answer the question.

SINGER: I tried last night in my talk to clarify some of the errors simply in historical fact which have now been repeated by Mrs. Simring. So, I would like to repeat that again. The two types of experiments which were the subject of the voluntary deferral that was requested by the Berg Committee in the summer of 1974 are still either prohibited or subject to the highest containment restrictions in the NIH guidelines. In fact, those conditions are not presently available, so that at the present time we still have the condition of the July 1974 moratorium in effect.

If the NIH had not proceeded to develop guidelines, then we would have simply that deferral in effect, and what we have instead are a lot of containment recommendations, a lot of requirements, and a very serious analysis of various other kinds of experiments. We have got to get those facts plain and clear, and we cannot talk about the lifting of a moratorium which never occurred.

SIMRING:  I don't think that Dr. Singer has addressed my main question, which is why did the program of evaluation of hazards not proceed as requested by the letter signers, as requested or mandated in the *Federal Register*?  Why was that not called for?  Why did all of you scientists accept the guidelines without the mandated necessity to assess the hazards beforehand?

RICH:  I think what Dr. Singer just said is that setting up the guide- lines allows the scientists to proceed with the evaluation, at the same time prohibiting those high-risk experiments.

SIMRING:  I don't feel that answers the question.

RICH:  It is impossible to get information unless one does experiments.

SIMRING:  Then the *Federal Register* should have been discussed openly and changed or canceled so that everybody could understand what was going on.  Instead nobody referred back to it at any time.

RICH:  Yes, Dr. Stetten?

DeWITT STETTEN, National Institutes of Health:  I am the Chairman of the Committee to which Mrs. Simring has referred, and I can assure those here and Mrs. Simring that attention has been given by the Committee to the conduct of experiments for the purpose of evaluating the limits of hazard.  If Mrs. Simring would trouble to read the minutes of our meetings which have occurred at intervals over the past two years, she could learn a good deal about this.  The experiments, unfortunately, are complicated, and they cannot be done instantan- eously.

In the first place, the P4 facilities required by these experi- ments are even now not fully completed.  These are complicated and expensive things to build, and we hope within the next few weeks to have a modest P4 facility on the grounds of the National Institutes of Health.

The very first experiments which will be conducted in this facili- ty will be performed at the request of the Committee by Dr. Wallace Rowe and his colleagues and were designed at a meeting of the Com- mittee at La Jolla and will follow these lines.  They will be designed to test the impact of polyoma virus DNA included in a suitable vector inserted into a host and administered by various routes to suscep- tible animals to ascertain whether this oncogenic virus is infective and is expressed under these circumstances.

We have had other experiments, and some of them have been conducted by people sitting in this room.  Dr. Roy Curtiss and his colleagues have been working continuously and hard under contract to the National Institutes of Health in the development of safer hosts for plasmids.  In addition, there are a number of other contractors of the NIH who have been developing and testing safer bacteriophage *E. coli* combinations.  We now have, I believe, five accredited EK2 combinations, that is, safer host-vector systems which were nonexistent two years ago.  I must present this as a significant accomplishment in the direction of the charge of the Committee.

SIMRING:  I would like to say, Dr. Stetten, that I do deeply appreciate the complexities of setting up an experimentation program.  But even so, they were mandated to precede the setting up of guidelines so that no matter how expensive, costly in time, energy, place, construction, the guidelines should have been put off then according to the mandated *Federal Register* item.

   As for the experiment that is going to be put up at Bethesda that you described, the polyoma recombinant DNA virus work, I understand, if I am not mistaken, that it is to take place in the highest-security containment laboratory possible, P4.  But it is located, I am told, and I have seen a picture of it, in a mobile trailer.

   Now, when some scientists wrote in to the NIH to protest this rather startling fact, they were informed that they really should not worry too much because it had a seven-foot cyclone fence around it with a lock.  This is in correspondence from the NIH that is available to anybody, and I suggest that in an unprecedentedly hazardous technology of this type that we proceed much more cautiously.

MEREDETH TURSHEN, Oil, Chemical and Atomic Workers International Union:
I would like to once again draw another analogy, since that seems to be our mode of communication this morning.  My analogy is to the long years of research that have been done on malnutrition and early childhood brain damage which I witnessed in my many years of work at the World Health Organization.  It seems to me that the promise of scientists that that research was justified by the fact that they were going to find a cure for malnutrition if they carried out all this esoteric research is not justified, and it seems to me that every mother, whether illiterate or with a Ph.D., whether in Mississippi or Tanzania or in the rice paddies of Bangladesh, knows what the cure for malnutrition is.  It is food.  It seems to me the kinds of things that have been said here this morning about the possible benefits of this research are benefits to the scientists who are carrying out the research.  It will benefit their careers.  It will give them lovely junkets to wonderful buildings like this one in Washington, and it will allow them to continue to do this research because as the research will be assessed five years from now, we will find we have not had enough research, and we will go on with more billions of dollars being granted whether by private industry or by the government if we have more research.

   My question is, what could conceivably justify recombinant DNA research, given what has already been said about the potential risks and given what we already know about public health and preventive approaches to disease and environmental deterioration?  We know what is in it for scientists, but what is in it for workers?

116

STEPHEN TOULMIN:  I think the question is a very important question about
   social priorities, which happens not to be part of our agenda.  I
   would have great sympathy with her if she were to raise the same
   question down at the Pentagon because if we are going to find
   sources of finance for improving the nutrition of young children
   in the world, which is a cause that none of us can fail to feel
   sympathy for, it seems to me that the defense budgets of all the
   nation states in the world are the first target to go after.  So far
   as I can see, the comparatively small part of national budgets which is
   allocated to scientific research of any kind is far from being the first
   obvious target for cutting.  It has been savaged enough in the last
   ten years already.

   I think that if political issues of this kind are going to be
   raised we could well direct our attention elsewhere, and I am even
   prepared to join in registering with the right kinds of protest so
   long as they have the right addresses on the envelope.

AUDIENCE:  For those of the audience who realize that a question that we
   should be addressing at this point in the national debate here is the
   ethical and the moral questions and the entry of corporations, I think
   maybe we should get together and discuss how we can get this issue
   to the public and get on with that and discuss the shortcomings of this
   Forum.  I would propose that we have a meeting tonight during the
   workshops that would be a caucus, not just a workshop on this issue,
   but a caucus with those concerns, that we determine this afternoon
   what size room we will need, that we conduct this on a town meeting
   kind of agenda and that if we have a large enough group of people
   maybe we should have it right in this room.

REFERENCES

1.  Kotelchuk, D.  1975.  Asbestos--science for sale.  Science for the
    People VII(5):8-16.
2.  Schwartz, C.  1976.  Scholars for dollars:  the business-government-
    university consulting network.  Science for the People VIII(1):4-9.
3.  Phillips, C., and S. Tafler.  1976.  In the land where Coca-Cola is
    queen.  Science for the People VIII(2):12-14.
4.  Harris, M.  The withering green revolution.  Nat. Hist., March 1973,
    pp. 20-23.
5.  Harris, M.  1977.  The political economy of health.  Radical Review
    of Political Economics, vol. 9, no. 1.  (Available from the Union of
    Radical Political Economists, 41 Union Square, West, Room 901, New
    York, NY  10003.)
6.  Commoner, B.  1972.  *The Closing Circle*.  Bantam paperback.  (See,
    for instance, chapter 4 on Los Angeles air.)
7.  Beckwith, J., and L. Miller.  Behind the mask of objective science.
    The Sciences, November/December 1976, p. 16.

CASE ANALYSIS 2
THE DANGERS OF PLANNED OR
INADVERTENT LABORATORY
INFECTIONS AND EPIDEMICS

# THE EPIDEMIOLOGY OF ENTERIC INFECTIONS AND LABORATORY INFECTIONS

H. Bruce Dull

*Assistant Director for Program, Center for Disease Control,*
*Public Health Service*

One can gather from my title that my plan is to survey data which, although not derived directly from recombinant DNA research per se, has
relevancy to the topic of this forum.  Besides the two items listed
in the title, however, I should like to comment on two others, which
from an operational point of view may also have some importance.  Thus
I intend to comment on the epidemiology of enteric infections, to review
our knowledge of laboratory-acquired or laboratory-associated infections,
and then to discuss our experience in the Center for Disease Control
with the importation and shipment of pathogenic agents and with the licensing of clinical laboratories under the Clinical Laboratory Improvement Act of 1967.
    The overall intent of my remarks is not so much either to reassure
or to alarm with respect to recombinant DNA research, but to support
several conclusions that I think are fairly evident even in prospect.
One is that there has been, is at the moment, and presumably always will
be a small, recognizable risk of infection to those working in or having
some association with laboratories where microbiological agents capable
of infecting human beings are being handled.  A second is that in the
regular activities of medical diagnosis and treatment, as well as in the

research mode or in the regulatory aspects of laboratory practices, there can be exposures to microorganisms that can lead to human disease. And a third is that from collective past experiences in laboratory containment, the best protection against infection risks is acknowledging that they can occur, defining them as well as we can, developing safe practices and good protective equipment, organizing surveillance to detect problems as soon as possible, making known whatever problems do occur so we can all profit from them, and continually updating all of our efforts to ensure safety.

EPIDEMIOLOGY OF INFECTION

First, let me tell you several relevant things about the epidemiology of infections in general, particularly enteric infections. Although I don't want to appear too much to be laboring basic epidemiologic principles, I should remind you first of all that there are three fundamentals in the spread of disease that essentially always apply. Obviously, first is the potentially infectious agent, the organism itself. The organism's unique characteristics help clarify the likelihood of its spread, not only the probably applicable routes of spread but the quantitative aspects of infectious dose and the like. Under the unusual environment of the laboratory, where organisms are found in far purer and perhaps more concentrated form than in nature, one must interpret and extend from the natural experiences, still recalling the characteristics of the organism itself.

The second item in the epidemiology of infection is the potential host, including the host's personal characteristics of age, sex, and general state of health. But, again, the host in the laboratory setting may have characteristics unlike the general population, not only with respect to personal factors but also to all-important behavioral ones.

Third, and critical in a laboratory setting, is the kind of exposure that a potential host has to a potential infectious organism. Exposure in a laboratory may have a remarkable range of varieties in its intensity, quality, and, indeed, in the specific events that could relate not only to routine procedures but accidents as well.

Before carrying our infection model further and using it as a backdrop for a general discussion of laboratory-associated infections, a few additional items need to be pointed out about the *process* of infection. These comments are primarily to remind you that organisms vary in their relative *infectivity*; that is, their ability to adapt to a human environment and to multiply there. They vary with respect to their *pathogenicity*; that is, the ability to create a clinical response that is observable as disease. They vary, of course, with respect to their *virulence*; that is, the severity of a pathologic process. And finally, they vary in *antigenicity*; that is, the ability to stimulate specific immunity. These biological variations are fundamental in considering laboratory-associated infections from the standpoint not only of the disability they may cause but also the chances for and ease of their detection.

On the one hand, there are chances, based on characteristics of the organism, of the host, and of the kinds of exposure, that one may become infected. And on the other, the infection produced may be mild or severe, may or may not result in clinical disease, and may leave immunologic evidence sufficient to connote infection or not.


ENTERIC BACTERIAL INFECTIONS

Applying some of these general concepts of infection to the epidemiology of enteric infections, several conclusions can be drawn. In distinction to diseases spread by the respiratory tract, those with an oral route of spread appear considerably more difficult to transmit. This relates largely to the infectivity of enteric bacteria in which, generally, a large multiplicity of organisms is needed. For example, commonly the dose of enteric bacteria needed to infect involves $10^7$ to $10^9$ organisms. In nature, therefore, grossly contaminated sources or a phase of "amplification" of enteric bacterial pathogens is needed to achieve a sufficiently large dose to be infective. Commonly, vehicles such as food or milk in which bacteria can multiply are implicated in human disease because in them a sufficiently high titer of organisms can be achieved to result in infection. I should point out, however, that there are a few enteric organisms, like shigella, which can infect with as few as 200 or 300 organisms.

Specifically with respect to *E. coli*, which like salmonella species requires large doses to infect, there are only a few examples of this organism's infecting healthy people. One example is the episode of a few years ago where at Crater Lake National Park the water supply was greatly contaminated with *E. coli*, and approximately 2,000 cases of human disease resulted.

Commonly, *E. coli* is no problem for healthy children and adults, and the only outbreaks occur in newborn nurseries in hospitals where the host susceptibility and physiologic immaturity do not provide the usual barriers to infection.

It appears that most enteric bacteria require large doses to infect because the human alimentary tract is a relatively hostile environment in view of its enzymatic and acid-base characteristics, likely to inactivate many enteric organisms. For example, cholera is exceedingly difficult to transmit orally unless the pH of the upper GI tract is sufficiently alkaline to protect this rather fragile organism. With this and many other enteric pathogens, characteristics of the GI tract are key determinants in the epidemiology of oral transmission of infection. By contrast, respiratory infection may occur with relatively few microorganisms--some epidemiologists feel that one infectious organism in the right place at the right time may be sufficient. It is for the reason of easy person-to-person spread that respiratory diseases may involve large numbers of people relatively easily, as is the case with influenza virus.

Again a note of caution must be entered. The laboratory may be, as we have noted, an unusual environment with respect to exposures. Enteric pathogens not normally infective could appear in sufficiently high titers to infect, where the same organism in nature might have little chance of survival. In some cases, laboratory exposure might introduce an enteric organism in an atypical way and create infection. Important in this regard, however, is the well-founded observation that secondary spread of enteric organisms, even if infection were acquired by some host in an unusual environment or under unusual circumstances of exposure, would relate to the characteristic patterns of an organism's epidemiology.

One other epidemiologic and clinical characteristic of enteric bacterial infection of some relevancy is the occasional development of a carrier state. Classic examples, of course, are some salmonella species, especially *S. typhi*. However, the risk of spread from persons who are chronic bacteria carriers is very small unless food handling or such activities plays a role in the process of transmission.

In general, then, the epidemiology of enteric bacterial diseases is characteristic in the limited person-to-person spread of these organisms largely as a result of their requiring a large dose and the proper conditions for exposure. Few secondary cases occur--largely related to the relatively well sanitized environment in which we live. And only when multiplication of bacteria occurs in an intermediate phase of the transmission process can enteric bacteria usually attain sufficient titers to be infective to healthy persons.


LABORATORY-ASSOCIATED DISEASE

Let me turn now to a brief survey of our knowledge of laboratory-associated disease. Most of this is from information reported in the medical literature, especially from a review by Dr. Robert Pike which covers published accounts of laboratory diseases in the nineteenth and twentieth centuries and a special survey of experiences through 1974.

It is important to recognize that, since there has never been a formal reporting mechanism for laboratory-associated infections, available data derive from events with inherent scientific intrigue or represent unusual cases or clusters of cases or highlight unusual circumstances of or atypical routes of infection. Data do not come at all from a systematic compilation of the experiences in laboratories. There obviously is underreporting, particularly of conditions that might be commonplace or difficult to detect. On the other hand, there may be some overreporting in the sense that interest in these illnesses may have identified some as having a laboratory source of exposure but which may, in fact, have been acquired in nature. This could be the case with diseases like hepatitis, typhoid fever, various viral respiratory diseases, and such. Such reports undoubtedly do not compensate for or offset the likely underreporting but do suggest that our data base on both sides is not solid.

As to the findings, Dr. Pike indicates that through 1974 approximately

4,000 cases of laboratory-associated infection were documented.  Case
fatality was about 4 percent.  Most of the deaths occurred prior to 1945
and were related to laboratory outbreaks and not to individual cases.
A common cause for such laboratory epidemics was a centrifuge-related
aerosol of the organisms of brucella or psittacosis with resulting exposure of a number of staff persons.  More recently laboratory infections
have represented single cases.

The etiologies of the 4,000 cases reported by Dr. Pike are interesting in that 43 percent of them were caused by bacteria and 27 percent by
viruses.  Of the bacterial diseases, enteric organisms accounted for only
22 percent but, importantly, were related to 35 percent of fatal bacterial infections.  The bacteria principally involved were typhoid,
other salmonella, shigella, and cholera.

The temporal trend in decreasing numbers of laboratory-associated
infections, especially with respect to bacterial infections, is remarkable.  Although viruses remain relatively important causes of infection,
in the last two to three decades there has been a tenfold decrease in
bacterial infections associated with the laboratory.

Of the laboratory-associated infections, only 20 percent can be related to laboratory accidents, the rest with the laboratory environment
in general.  Only about 20 percent of specific infections occurred in
persons who were actually working with the agent that caused the disease.  Thus, the great majority of infected persons had no definitive
contacts other than with the laboratory environment in general.

An important finding in Dr. Pike's research was that more than 50
percent of the laboratory-associated infections occurred in research
laboratories and 17 percent in diagnostic laboratories.  Before you
conclude, however, that research therefore is particularly hazardous,
remember that oftentimes the diagnostic laboratory usually deals with
a "routine spectrum" of pathogens while a research laboratory, by
virtue of the kind of research being done, may encounter organisms which
are inherently more dangerous or hazardous.

With respect to the sponsorship and not the purpose or function of
the laboratory, Dr. Pike found that infection was more related to the
agents being used than to either the sponsorship or the purpose of the
work.  Many of the laboratories reporting infections were government
sponsored and others were in private institutions, hospitals, universities,
or colleges.

Notable in Dr. Pike's survey, and especially germane to our inquiry,
is the temporal decline in the number and severity of laboratory-
associated infections.  Why has this occurred?  The answer would appear
to be fairly obvious, particularly with respect to bacterial agents.
There clearly have been a greatly increased awareness of laboratory
hazards, improved containment techniques and equipment, and the regularized
use of safety devices.  Furthermore, many of the infectious agents that
were particularly problematic, such as those causing brucellosis, Q fever,
and psittacosis, are little studied today, at least without good containment facilities.

## SHIPMENT OF ETIOLOGIC AGENTS

Let us now turn to the surveillance of pathogenic agents in commerce and to their supervision.  In this regard I would like to speak particularly about the responsibility which CDC assumes for the Public Health Service in granting permits for the importation and interstate shipment of etiologic agents.  By etiologic agents, I mean microorganisms or toxins capable of causing *human* disease and not biologicals or diagnostic specimens.  The importance in reviewing control over the shipment of etiologic agents is that it may contain some clues to the surveillance and management of recombinant DNA organisms.

The key elements in the supervision and control of potential hazard in the importation and interstate shipment of etiologic agents is in having advance knowledge of which agents are to be transported and what containment procedures would be suitable.  Once aware of the agents intended for transport, their relative hazard in terms of standard classification and their containment in packaging in accordance with this hazard are controlling elements.  Included in the packaging standards are labeling and administrative considerations, including instructions on handling damaged packages or those that cannot be delivered.

In the Center's management of this activity, several hundred applications for importing etiologic agents are reviewed each year.  In 1976, 353 permits were issued.  During the years that CDC has supervised this effort, there have been no reported incidences of untoward events or of any hazards to human health.


## CLINICAL LABORATORY LICENSURE

The final subject I would like briefly to review is CDC's experience under the Clinical Laboratory Improvement Act of 1967, a statute intended to enhance and sustain the quality of clinical laboratory performance in the United States.  The basic methodology of the Act was to grant federal licenses to laboratories involved in interstate commerce unless equivalent state license or other quality control certification existed.  The basis for granting licenses were staff qualifications, facilities review, and demonstrated proficiency.

In the seven years since CDC assumed responsibility for regulating and licensing clinical laboratories subject to the federal law, some 2,000 license applications have been reviewed in a number of categorical areas such as bacteriology, chemistry, hematology, cytogenetics, and others.  As part of the application review, records of some 29,000 staff persons have also been evaluated.  From review of license or license update applications and inspection of facilities, several important conclusions can be made.  If one wants to be assured of quality performance in laboratory practice, one needs to look at three elements.  One is staff qualifications, including the depth and breadth of training and experience. Probably more than anything else, characteristics of laboratory staff have been the best predictive determinants of quality.  Second is the regular pursuit of internal quality control programs, with efforts to standardize procedures and carefully supervise functions. And third is laboratory involvement in a continuing proficiency testing program to determine personnel competency and evaluate performance.

Currently there are 1,036 U.S. laboratories with federal licenses. This is only about 7 percent of the estimated 15,000 clinical laboratories in existence not counting the small laboratories in physicians' offices. Some of the laboratories without federal licenses have state or other licenses or other certification attesting to performance. Nevertheless, clinical laboratories, many of which process microbiological agents, are widespread and do represent environments in which laboratory-associated infections could occur and in which surveillance could be conducted.

One final comment from the Center's experience with licensure involves surveillance of bacteriological proficiency. In the seven years the Center has been involved with licensing, there has been an overall improvement to 80-85 percent accuracy in the ability of licensed laboratories to diagnose bacteria. Before one concludes that this is not particularly good performance, let me indicate that the proficiency testing from which these data derive is done with an eye toward making correct answers difficult to obtain. In terms of routine diagnosis, laboratory performance is probably far better. One example of this is in a recent survey of the ability of licensed laboratories to diagnose E. coli when it was presented as a single agent in a battery of specimens. Under these conditions, 98 percent of laboratories properly identified the bacterium. On the other hand, although laboratories identified 80-85 percent of bacteria correctly in general, some organisms were identified properly only 40-50 percent of the time. There is still room for improvement.

In summary let me reiterate the observation I made at the outset. Although experience with recombinant DNA organisms and the risks they might theoretically pose is essentially nonexistent, one can derive considerable information from surveillance of laboratory-associated infections in general. In this context it is clear that laboratories handling microbiological agents have and presumably will always have a small risk of infecting those encountering the laboratory environment. The risks are small, and they can be minimized or made essentially negligible by adequately preparing the staff to deal with the potential risks, by instituting and maintaining good containment practices, and by developing ongoing surveillance activities. The best preparation for assuring safe performance is to admit the existence of some risk and to prepare physically and intellectually to see that any potential problem is anticipated and minimized.

# EPIDEMIOLOGICAL AND EVOLUTIONARY ASPECTS OF RESEARCH ON RECOMBINANT DNA

## Bernard D. Davis

*Adele Lehman Professor of Bacterial Physiology, Harvard Medical School*

Several charges have been leveled against proponents of research on recombinant DNA: selfishness, in risking the production of an Andromeda strain in order to satisfy their curiosity; blasphemy, in meddling with evolution; and irresponsibility, in bringing us closer to genetic engineering in man. These charges have been based on the assumption that we are entirely in the dark in trying to assess these dangers. But this is not so. On the question of the hazard of an epidemic a good deal of pertinent theoretical and factual information is available from the science of epidemiology (concerned with the genetic and the ecological factors that influence the spread of disease), and from evolutionary theory (of which epidemiology may be viewed as an applied branch). Evolutionary theory also has serious implications for the more long-range danger of possibly fouling up evolution. This paper will review some of the relevant information, concentrating on the risk of producing an epidemic, and considering this problem in terms of three component risks: that a harmful organism may inadvertently be produced, that it may cause a laboratory infection, and that it may spread into the community.

In approaching the subject from this perspective, I would like to express my agreement with Jonathan King on one point: that the Asilomar Conference did not have sufficient input from experts in infectious disease. I further regret that this field continues to be relatively neglected in the current discussion. For since we are dealing more with a problem in epidemiology than with one in molecular biology, epidemiological principles provide the most reasonable basis for present estimates of risk. Moreover, though the risk of an epidemic will ultimately have to be assessed in terms of future experience with various recombinants, even the most favorable experience will not eliminate the specter of a future Andromeda strain unless we interpret it in terms of epidemiological principles.

## UNDERLYING PRINCIPLES

### Natural Selection

Evolutionary change arises ultimately from hereditary variation, but its direction is dominated by natural selection. It is dramatic for George Wald to state that research with recombinants is dangerous because "a living organism is forever"--but a more balanced statement would also note that only an infinitesimal fraction of the products of evolutionary experimentation survive, the rest being ruthlessly culled out by natural selection. In particular, within a species the process of sexual reproduction produces a virtually infinite variety of recombinants, among which the standard pattern of selection is a stabilizing (normalizing) one: excessive deviations from the norm make an organism less effective in the Darwinian competition. It is only when the environment is altered that certain deviants from the norm turn out to be better adapted to the new environment, and selection then becomes directional.

It should also be emphasized that *all natural selection is for a balanced genome*. A gene that increases or decreases a trait is selected for, not in a vacuum, but only if it is coadapted to the rest of the organism's total set of genes.

The Meaning of Species

As evolution proceeded from prokaryotes (bacteria with a single chromosome) to eukaryotes (higher organisms, with a more complex genetic apparatus), it created the mechanism of sexual reproduction. By reassorting the genes of paired parents this process provides vastly increased genetic diversity for natural selection to act on. But since a successful organism must have a reasonably balanced set of genes, the production of unlimited recombinations from the total pool of genetic material in the living world would not be useful. Hence the development of sexual reproduction was accompanied by the development of species: groups of organisms that reproduce in nature only by mating with other members of the same group, and not with members of other species. The evolutionary value of such fertility barriers between species is clear: to avoid useless production of grossly unfit, nonviable progeny.

Bacterial Genetics

Though Darwin was unaware of the existence of the invisible world of microbes, their slow absorption into the Darwinian framework began, unwittingly, with Pasteur's demonstration that different media, such as milk or grape juice, select for different organisms from the same mixture of contaminants that can reach them from the air. But it was not until the 1940s that heredity in bacteria was shown to depend, as in higher organisms, on unitary genes, linked on a chromosome and capable of mutation, transfer, and recombination. Indeed, with this development it became possible to use microbes to demonstrate the force of natural selection in an overnight experiment. In addition, with the emergence of molecular genetics from microbial genetics it became possible to provide the ultimate proof, from DNA sequences, for a crucial prediction of modern evolutionary theory: that the accumulation of changes in genes is the basis for the divergence of organisms in evolution.

Unlike eukaryotes, prokaryotes ordinarily reproduce by asexual cell division, which means that the genetic properties of a strain remain constant for generation after generation, except for rare mutations or for rare transfers of a block of genes from one cell to another. These gene transfers, which are usually mediated by plasmids or viruses, do not show a sharp species boundary: they simply become less efficient the greater the evolutionary separation between the donor and the recipient. Prokaryotes therefore have no true species. *E. coli*, for example, is the name given to a range of strains with certain common features and also with a variety of differences--in surface molecules, nutrition, growth rate, sensitivity to inhibitors, etc. These differences determine the relative Darwinian fitness of various strains for various environments.

## Bacterial Ecology

Every living species is adapted to a given range of habitats. The set of bacterial strains called *E. coli*, and such closely related pathogens as the typhoid and the dysentery bacilli, thrive only in the vertebrate gut. In water they survive temporarily but quickly die out. (Indeed, for that reason the *E. coli* count of a pond is a reliable index of its continuing fecal contamination.) In the gut there is intense Darwinian competition between strains, depending on such variables as growth rate, nutritional requirements, ability to scavenge limited food supplies, adherence to the gut lining, and resistance to antimicrobial factors in the host. Hence most novel strains are quickly extinguished, in the kind of competition envisaged by Darwin for higher organisms. With bacteria the process is very rapid, because the generation time is as short as twenty minutes and the selection pressures are often intense.

It is easy to demonstrate that the environment in the gut (i.e., type of food and physiological state) plays a decisive role in determining the distribution of organisms in its normal flora. For example, when a baby shifts from breast feeding to solid food the character of the stool changes dramatically, as lactic acid bacteria, which produce sweet-smelling products, are replaced by *E. coli* and other foul organisms. Moreover, efforts to reverse the process in adults, by administering large numbers of lactic acid bacteria in the form of yogurt, have not been successful.

## Pathogenicity

Various kinds of infectious bacteria differ from each other in several distinct respects: infectivity (i.e., the infectious dose, ranging from a few cells of the tularemia bacillus to around $10^6$ cells of the cholera vibrio); specific distribution of the organisms in the body; virulence (i.e., the severity of the disease once the infection has overcome natural resistance); and communicability from one individual to another (including length of survival in nature). Each of these attributes, like any complex property, depends on the coordinate, balanced activity of many genes, capable of independent variation.

It is especially important to distinguish the ability to produce a serious disease from the ability to spread. For example, the tetanus bacillus produces a powerful toxin, but it is a normal, noninvasive inhabitant of the gut: it can cause fatal illness only when it gains access (usually by trauma) to a susceptible tissue, and so a patient with tetanus is not a menace to his contacts.

ESTIMATION OF THE HAZARDS

In turning now to the risks, I would note that they are often not as directly commensurable with benefits (i.e., expressible in similar units) as are costs compared with benefits. For this reason a particular risk must be judged for acceptability not only in terms of a comparison with benefits but also in terms of its probably increment to the related risks that we already live with. I would further emphasize that it is easy to draw up scary hypothetical scenarios if one's imagination need not be limited by considerations of probability. But any realistic discussion must consider probabilities. And as I mentioned earlier, we must consider three probabilities: that experiments with a given kind of DNA will produce a dangerous organism, that that organism will infect a laboratory worker, and that the organism will escape and spread in the community or the environment.

Risk of Producing a Harmful Organism

There is no doubt that molecular recombination *in vitro* could produce pathogenic derivatives of *E. coli*. For example, if a strain carrying the gene for a potent bacterial toxin multiplied enough in the host, or even if it could not multiply but were taken up in a large enough dose, it could cause disease. A strain carrying a tumor virus might also be hazardous. However, its production of a pathogenic effect is less certain. For unlike a toxin producer, such strains would require for pathogenicity more than the normal function of the foreign DNA within the bacterial carrier: it would require release of that DNA from the bacterial cell and its infection of animal host cells. While that probability may be very low, we cannot assume that it is negligible. Both these kinds of strains are appropriately prohibited in the NIH guidelines today.

I would like to concentrate on a kind of experiment that is allowed, but that is causing great concern and is restricted to P3 facilities: the so-called "shotgun" experiment, in which one transfers random fragments of DNA from mammalian cells. Two considerations convince me that the danger in such experiments has been enormously exaggerated. First, such cells have a million gene equivalents, and since each recombinant strain would contain only a few genes, the probability of isolating a strain with genes for a toxic product or for a tumor virus is exceedingly low.  Second, I would seriously question whether the novelty that we fear in the products of such experiments is real.

The reasons for this doubt are the following.  It is known that bacteria can take up naked DNA from solution.  In fact, two different strains of pneumococcus have been shown to be able to produce a third, recombinant strain in an animal body, by release of DNA from a lysed cell of one strain and its uptake by an intact cell of the other.  Moreover, in the gut, bacteria are constantly exposed to fragments of host DNA, released by death of the cells lining the gut; while bacteria growing in carcasses have a veritable feast.  To be sure, the efficiency of uptake of DNA by bacteria (especially the kinds found in the gut) is very low; but on the other hand, the scale of the exposure in nature is extraordinarily large--around $10^{20}$ bacteria are excreted collectively by the human species per day.  Hence it seems virtually certain that recombinants of this general class have been formed innumerable times over millions of years and are being formed in nature today.  If they had high survival value we would be recognizing short stretches of mammalian DNA in *E. coli*.  We do not.  On the other hand, naturally occurring recombinants might be appearing and even causing transient epidemics, which are escaping our attention.  But then we would have to ask how much our laboratories could add, performing experiments on the scale of $10^{10}$-$10^{12}$ bacteria.


Risk of Laboratory Infection

Having considered the probability of inadvertently producing a harmful organism, we must now consider the probability that such an organism would cause a laboratory infection.  Let us assume the worst case, at present prohibited:  an *E. coli* strain producing a potent toxin absorbable from the gut, such as botulinus toxin.  The danger of harm from a laboratory infection with such a strain would be real.  However, there are a number of reasons to expect it to be less than the danger encountered with the pathogens that are handled every day in medical laboratories:

1.  In the history of microbiology about 6,000 instances of laboratory infection have been recorded.  Moreover, these cases were largely due to various agents of respiratory infection, spread by droplets; and the rate has dropped markedly since safety cabinets were introduced in the 1940s.  In contrast to such respiratory infections, enteric infections arise through the swallowing of contaminated food or other material. Hence even the most virulent enteric pathogens are relatively safe to handle with simple precautions, such as not eating or smoking in the laboratory.

2.  Strain K-12, used in almost all genetic work with *E. coli* (including current work with recombinant DNA), has been transferred in the laboratory for over fifty years, and during this time it has become well adapted to artificial media, at the cost of becoming de-adapted to the

human gut. In fact, in recent tests in man this strain disappeared from the stools within a few days after a large dose (much larger than what one would expect from a laboratory accident). Its problems of survival outside the laboratory are analogous to those of a delicate hothouse plant thrown out to compete with the weeds in a field.

3. The addition of a block of foreign DNA to an organism will ordinarily decrease its adaptation to survival in nature. The contrary likelihood, of improving adaptation by such an insertion, is obviously all the smaller if the source of the DNA is distant in evolution from the recipient. A pertinent analogy here would be that of taking a specialized part from one kind of machine (e.g., an automobile) and expecting it to work well in a very different machine (e.g., a watch).

4. A very large safety factor is added by the provision in the present guidelines for biological containment. All work with mammalian DNA must be carried out in EK2 strains, which have a drastically impaired ability to multiply, or to transfer their plasmid, except under very special conditions provided in the laboratory. The presently certified EK2 strain has several stable mutational defects (i.e., deletions) that prevent it from multiplying under the nutritional conditions of the gut. But the protection goes much further, and reaches a degree that is unprecedented in the annals of man's exploration of potentially hazardous new materials: this material has been coded for self-destruction. For example, these mutant cells require diaminopimelate, a constituent of cell wall; and without it they can continue to grow and expand but cannot form more wall, and so they quickly burst. Accordingly, under conditions similar to those in the gut such an EK2 strain not only fails to multiply, but less than 1 in $10^8$ cells survives after twenty-four hours--and it would be an extraordinarily sloppy laboratory accident that would result in ingestion of as many as $10^8$ cells. In addition, while the cells are dying off in the absence of diaminopimelate they are severely impaired in their ability to transfer plasmids to other, well-adapted cells--and this is the important point for the danger of spreading harmful genes. Finally, not only the cells but also the plasmids being used to carry recombinant genes are also weakened mutant derivatives, selected for severe impairment of their ability to be transmitted from the host cell to another cell.

We thus see that, even with a strain known to carry the gene for a potent toxin, the production of disease in a laboratory worker would require the compounding of two low probabilities: that the strain will initiate an infection and that it will survive long enough to cause harm despite its several disadvantages--that of being a laboratory-adapted strain, that of carrying the burden of foreign DNA, and that of carrying the very large burden of being a suicidal EK2 strain. With shotgun experiments we have a third, very low probability, already mentioned: that of having picked up a dangerous gene from normal mammalian tissue.

I conclude that with the kinds of recombinants now permitted the danger of a significant laboratory infection is vanishingly small compared with the dangers encountered every day by medical microbiologists

working with virulent pathogens.  And such dangers must ultimately
be balanced against the potential benefits.  In the United States, up
to 1961, of the 2,400 recorded cases of laboratory infections, 107 were
fatal--over half of these from diagnostic laboratories.  Balancing this
cost, millions of lives have undoubtedly been saved by bacteriological
research and diagnosis.

On the other hand, even if the risks in recombinant DNA research are
really small, it is important to keep all the probabilities low.  Hence
it is important for molecular biologists working in this area to learn,
and to use, the standard techniques of medical microbiology.  Indeed,
the main benefit from the current discussion might well be the enforcement of such practices.

Risk of Spread

I now come to the most important point of all from the point of view of
the public:  the enormous difference between the danger of causing a
laboratory infection and the further danger of unleashing an epidemic.
Let us look at a few facts.  In our government's bacteriological warfare
laboratories at Fort Detrick, working for twenty-five years on the most
communicable and virulent pathogens known, 423 laboratory infections were
seen.  Moreover, most of these infections occurred via respiratory
transmission, over which control is very imperfect.  Nevertheless, only
a single probable case of secondary spread to a member of the family
or to any person outside the laboratory was seen.  Similarly, in the
Center for Disease Control of the U.S. Public Health Service, 150
laboratory infections were recorded, with only one case of transmission
to a family member.  Elsewhere in the world about two dozen laboratory-
based microepidemics have been recorded--and each involved at most a
few outsiders.

With enteric pathogens the danger of secondary cases is minimal, for
with this class of agents modern sanitation provides infinitely better
control than we can provide for respiratory infection:  the appearance
of a case of typhoid, in contrast to that of influenza, does not lead to
an epidemic.  Enteric epidemics appear only when sanitation is poor or
has broken down, or when a symptom-free carrier with filthy personal
habits serves as a food handler; and such epidemics are always small
(except when sewage freely enters the water supply).  Moreover, the focus
of some critics on the debilitated or the young, as exceptionally susceptible victims, is not realistic:  we are dealing with interruption
of the chain of transmission, and not with wide spread of the organisms
at a low density.

This information is clearly pertinent to recombinants in *E. coli*.
For while widespread apprehension has arisen from the presumption that
this procedure will produce biparental chimeras, with totally unknown
properties, the fact is that the recombinants envisaged are all genetically 99.9 percent *E. coli,* with about 0.1 percent foreign DNA added.  It
is not conceivable that such an organism could have a radically

expanded habitat, no longer confined to the gut.  It is even harder to
see that the organism would be more communicable, or more virulent, than
our worst enteric pathogens, which cause typhoid and dysentery.  The
Andromeda strain remains entertaining science fiction.

I conclude that if by remote chance a recombinant strain should be
pathogenic, and if it (or a recipient of its plasmid) should cause a
laboratory infection, that infection would give an early warning.  Moreover, if a case should appear outside the laboratory, the enteric habitat
of *E. coli,* combined with modern sanitation, provides powerful protection against the chain of transmission required for an epidemic.

Tumor viruses present a special problem.  Unlike other viruses, they
do not cause disease regularly after infection but require special circumstances.  Indeed, it is their occasional presence in apparently
normal animal tissues that has given rise to fear of shotgun experiments.

On the other hand, any conceivable infection by a bacterium containing a tumor virus genome would have a long latent period before disease
could appear, and so we would lack the early warning that would be seen
with a bacterium producing a potent toxin.  However, this loss of one
protective feature is balanced by the fact that viruses, by definition,
have their own means of spread.  Indeed, in general the natural spread
of viruses is even more effective than that of bacteria, for each infected animal cell produces thousands of infectious virus particles,
while each bacterium produces two daughter cells.  Moreover, since viral
DNA in a bacterium would have to get out of its host cell and get into
human cells, through an extremely inefficient process, it is hard to imagine that that DNA in a bacterium would be more hazardous than that same
DNA in its own infectious, viral coat, adapted by evolution for entering
animal cells.  Indeed, if we fear the danger of such indirect uptake of
unrecognized tumor virus DNA from normal mammalian tissue, via a bacterial
vector, we must ask whether the direct ingestion of such mammalian tissue, as in a "rare" steak, may not present at least as great a danger.
Finally, if we fear that tumor viruses are sufficiently widespread to
create a significant danger of being included in DNA fragments from
normal tissue, we must ask how much that wide distribution could be increased by the remote chance of inadvertent further spread by the bacterial hybrids created by shotgun experiments.

I am not suggesting that we should be concerned about the danger of
acquiring a cancer by eating rare meat (or by receiving a transfusion,
which inevitably has a fair chance of coming from a person with an un-
detected early cancer).  I am suggesting only that the danger of using
recombinant DNA to study tumor viruses must be judged against that background, as well as against the background of the virus's own distribution and inherent ability to spread.

In the light of all these considerations, we must ask whether the
danger of an epidemic really merits deep concern by the general public.
To be sure, the problem of minimizing the risk of laboratory infections
should concern those involved with such laboratories, just as with
laboratories dealing with known pathogens.  And I believe investigators

have the right to take such risks for themselves, as they do daily in working with pathogens (including such unknowns as the agent of "Legionnaires' Disease"). But we have seen that by any reasonable analysis the risk of producing a serious epidemic with *E. coli* containing random fragments of mammalian DNA seems very much less than the risk from pathogens that are being cultivated in laboratories all the time. I therefore see no realistic basis for public anxiety over this issue, any more than over the way laboratory work on known pathogens is conducted.


The NIH Guidelines

In the face of the alleged dangers that have been so vividly portrayed, I cannot blame the public for having a high level of anxiety. I also would regard the present guidelines as a reasonable response to that anxiety. On the other hand, in the light of the technical realities that I have discussed above, I would regard these guidelines as excessively conservative. This is especially true of the experiments with mammalian DNA, which offer enormous promise in the analysis of the structure and the regulation of mammalian genes and in the manufacture of valuable human gene products.

The guidelines contain a provision for periodic revision; and since these revisions (or the nature of any future legislation) will depend on public attitudes as well as on the results of actual experience with the organisms, there is need for a great deal of public education, based on the relevant scientific facts and principles.

In this connection I would criticize the *New York Times* for the article by L. Cavalieri on recombinant DNA in its "Sunday Magazine" (August 1976). Though the writer is a molecular biologist whose official credentials would lead the reader to expect a reasonable degree of objectivity, the article was inflammatory and it exhibited extraordinarily little understanding of either microbiology or evolution. In discussing *E. coli* as though it were a standard, uniformly distributed organism, which would carry with it through the world any additional genes that one might insert, the writer ignored the most important factor of all: natural selection among the innumerable strains of *E. coli*. He also made the remarkable statement that the insertion of tumor viruses into bacteria may make them infectious--as though viruses are not infectious. And he suggested that scientists working in this field may produce yet another Andromeda strain--as though the first strain existed in fact rather than in fancy.

Given the present level of public anxiety, scientists in this field seem quite willing to accept the guidelines. But I hope it will not be too long before these rules are modified in the light of further experience. For since the technique is potentially useful for a wide variety of problems, a requirement for excessively elaborate facilities will add up to a very large expense and will inevitably inhibit desirable experiments. The principle of erring on the side of caution is laudable up to a point--but if it is pushed too far it can end up being paralytic.

# INTERVENTION IN EVOLUTION

## The Prokaryote-Eukaryote Barrier

The hazard that we have been discussing--that of creating novel, danger-ous organisms--is a legitimate cause for public concern: there is no question about society's right to limit activities that may harm others. However, when we ask, with Dr. Sinsheimer, whether our increasing power to manipulate genetic material creates long-term evolutionary dangers, we are in quite a different area, involving the concept of dangerous knowledge rather than dangerous actions. Perhaps we can clarify the issue by trying to translate into more specific terms some of the gen-eral sources of apprehension that Dr. Sinsheimer has expressed in various publications.

1. He questions our moral right to breach the barrier between pro-karyotes and eukaryotes, since we simply cannot foresee the consequences. This argument seems to turn evolutionary principles through 180 degrees. Evolution is concerned with selection for fitness, in the Darwinian sense. The barriers that it has established between species are designed to avoid wasteful matings, i.e., matings whose products would be mon-strosities, unable to survive, rather than monsters, able to take over. Since survival of an organism depends upon a balanced genome, it is not surprising that evolution proceeds in small steps, which will not excessively unbalance the genome in one respect while improving its adaptation in another. And since for this reason even closely related species cannot form hybrids in nature, it is exceedingly unlikely that artificial transfers of genes between the most distant organisms--man and prokaryotes--would pass the test of Darwinian fitness.

2. "This is the beginning of synthetic biology." I wonder whether this statement can really be defended. Man has been meddling with evolu-tion since neolithic times, domesticating animals and plants by selective breeding, and also cloning and grafting plants.

3. "We no longer have the absolute right of free inquiry." But we never had: visibly dangerous procedures have always been subject to social limitations. But to invoke dimly foreseen, undefined dangers as a basis for limitation seems to be starting on the slippery slope of excluding dangerous ideas rather than dangerous actions.

4. A further push in this direction may be seen in the statement that power over nucleic acids, as over the atomic nucleus, "might drive us too swiftly toward some unseen chasm.... We should not thrust inquiry too far beyond our perception of its consequences." I would paraphrase this statement and suggest that we should not thrust our limitations on research too far beyond our perception of its hazards. Otherwise we will find ourselves reenacting the drama of Galileo and Urban VIII, and we will be trying to play the role of God (or of his representative). The analogy is uncomfortably close: for the mystical quality of the current argument suggests that at its core the issue is whether man's possible interference with evolution is not blasphemous.

Genetic Engineering in Man

Perhaps the most significant of Sinsheimer's statements is his suggestion that the study of recombinant DNA in bacteria is the beginning of
a genetic engineering that will ultimately extend to man.  Here, in
contrast to the vagueness of the preceding propositions, we finally
come to something concrete that one can wrestle with.

I would suggest that concern over genetic engineering in man is utterly irrelevant to the question of the danger of creating an epidemic;
hence it is irrelevant to Sinsheimer's recommendation that all research
on recombinant DNA be presently restricted to a few maximum-security
federal facilities.  This concern also seems irrelevant to the question
of breaching the prokaryote-eukaryote barrier; for while gene transfers
across this border at the cellular level, in either direction, are of
great scientific interest, it is hard to envisage any reason to try
to introduce into man genetic material from the opposite end of the
evolutionary spectrum.  Yet vague concern over possible extensions of
gene manipulation to man, even more than concern over epidemics or over
meddling with evolution in general, may lie at the heart of much of the
uneasiness over recombinant DNA research.  And because of the enormous
publicity given to our new power to splice blocks of DNA into plasmids,
we have perhaps lost sight of the fact that this development is no more
radical a step toward genetic engineering in man than are many other
steps, which have aroused no such public terror.  These include the isolation of a gene, its chemical synthesis, the cultivation of human cells,
the use of viruses to incorporate genes into those cells, and the achievement of genetic recombination *in vitro* between human cells and other
animal cells.

The prospects of genetic engineering in man received extensive discussion in 1970, which then subsided; and I see no reason to modify today
the analysis that I published then (Science *170*:1279), except to agree
with Dr. Baltimore that replacement of bone marrow cells may no longer
be very distant.  However, since the question has been reactivated by
the very different question of genetic engineering in bacteria, I would
like to make a few brief points.

First, as far ahead as it is profitable to look, the medical aim of
genetic engineering in man is simply gene therapy, for diseases due to
defects in single genes with a well-defined chemistry.  (Cloning is
another matter:  its specific aim is to avoid genetic recombination,
and its social purpose would not be medical.)  For gene therapy of most
hereditary diseases we would have to be able to introduce DNA in a reliable, controlled way, in the right cells:  and I believe we are still
a long way from that goal.  But even if this guess is wrong, and if we
succeed in genetically curing such diseases as phenylketonuria and
cystic fibrosis, it is clear that we would still be very far from being
able to manipulate in any useful way the large number of genes, all still
undefined, that specifically direct the development and the function of
the brain.  Moreover, in a developed organism, with an already formed
brain, no conceivable manipulation of DNA could reorganize the wiring

diagram of that brain--which is surely the main basis for the genetic component of human behavioral diversity. Hence the possibility that a tyrant could use genetic engineering to manipulate personalities seems still too remote to justify present concern. Finally, even if we could use genetic technology in this way, I would question whether the technological imperative would necessarily (or even likely) lead us to do so. For the simple but effective techniques of selective breeding and artificial insemination are already available, and yet they are not being used to influence the human gene pool.

Philosophical questions about the effects of science and technology on man's fate go back to Galileo--and the history of Italy's fate, in losing that early head start, should give us pause. For better or worse, we cannot unlearn the scientific method; and if we restrict it in one country it will turn up in another. To be sure, our world has only recently come to realize how large (and often unexpected) is the price for various aspects of technology, how finite our terrestrial resources, and how clumsy our responses to the need to limit the size of our population and its demands on those resources. Faced with these crushing problems, it is only too easy to take the benefits of science and technology for granted and to object to the new problems that they are raising. But in the long run it is difficult to see how we can plot a more prudent course than to continue to advance knowledge, while increasing our efforts to recognize (and to minimize) the hazards and the costs of its specific possible applications as soon as they become visible.

I share Sinsheimer's concern for the future, and his passionate advocacy of vigilance. But the vigilance must be directed at specific, definable applications of knowledge. Vigilance concerning new knowledge that might someday be misused is a threat to freedom of inquiry, and I believe a threat to human welfare. We may conceivably be entering dangerous territory in exploring recombinant DNA--but we are surely entering dangerous territory if we start to limit this exploration on the basis of our incapacity to foresee its consequences.

# DISCUSSION

DOROTHY ZINBERG, Program for Science and International Affairs, Harvard University: My question really comes out of some of the things which I feel that we have been missing. As a sociologist sitting and listening to this I feel that I have been a participant in essentially a political discussion, and yet nobody has really addressed the question of politics. I would like, as a teacher, to be able to go back with some information for my students.

I have recently had a very gifted student, a biochemist, who has just applied to graduate school, who was interviewed on his social distance from the Boston recombinant DNA group, and he turned out to be a totally apolitical young man and came out with flying colors. I, however, was very concerned about the implications of this interview.

The rumors I hear from a number of students and from junior faculty are that grants are really having an additional dimension added to them in their evaluation, which is that one's position on whether recombinant DNA research should go ahead is used as a variable in determining whether or not one gets one's grant. I would like to know whether the people here can talk to this. I say this not as an adversary, nor as Professor Sinsheimer said this morning, orthogonally, I hope, but really pedagogically, so that in this kind of arena we can talk about whether this is a valid concern, or whether or not the students and the junior faculty are unduly upset about what they are reporting as fact.

RICH: Dr. Stetten, would you care to comment on that? I don't know that anyone here has any insight in it.

STETTEN: I am not sure I can add anything, being surrounded by a nine-foot-high Anchor fence, which keeps this kind of rumor out. I am not aware in the grants process that questions in this area are being considered, because I cannot know what is going on in the minds of the study section members. I simply have no information on this at all.

WALD: If I may say a word to Dr. Zinberg, it hardly matters how much concrete reality one can involve the granting agencies in in this sort of question. The conviction is widely distributed among young scientists and people about to get their degrees and nontenured young faculty that if one ever expects a job or if one ever is to expect support from the granting agencies, or continued support from NIH, it is best to shut up about this. As to the effectiveness of this, I can only tell the people here that there came a crucial time after many meetings of our Biology Department at Harvard on this issue, when the junior faculty was really carrying the ball in opposition to the recombinant DNA, specifically the planting of a P3 facility on our fourth floor, when finally it came to a meeting with the dean. And when the dean sat down at that meeting he looked around and said, "Where is the Biology Department?" And indeed, we had heard the last of any talk in this direction from our junior faculty. So it hardly matters whether this becomes concrete as many other such things at the top echelons of NIH. This conviction is very widespread and very effective.

BERNARD DAVIS: When Dr. Wald says that conviction is very effective, it seems to me an action is effective, and that a conviction can only be effective if it is reflecting somebody's action. Therefore, it is important to know whether somebody actually is acting to interfere with people's getting grants on the basis of their opinions on this matter. The fact that a great many people believe that somebody is a Communist or somebody is something else is rather different from the question of whether a person is or is not something.

WALD: The action in this case is the silence of junior and nontenured faculty people, and young scientists.

I have been examining this situation for a long time. It may surprise some of those present that at the Marine Biological Laboratory in Woods Hole last summer this issue, which is beginning to draw the attention of the whole world, and in spite of the fact that it was suggested to the director that a high-level discussion occur there, this issue never came up in my experience, not only not in the form of a public discussion, but also not in the form of even private conversation. It just didn't come up. Interpret that.

BROWN: I have a statement to that question which was just asked about interference with grants. It was a study section which was held last year in which one member of the study section got up and said, "If there is any grant which has recombinant DNA in it, I will automatically vote to disqualify that grant."

JONATHAN KING: I would like to say that I think Dr. Zinberg has picked up a very sensitive issue, and I do hope someone keeps an eye on this very closely, about who does get funding and who doesn't. I know I personally would like to speak more about this, but I know my students are upset about it. There is no doubt about it. They get upset every time I appear in a public forum on this. They are worried about their jobs. They will say it to me. They won't say it publicly because they feel that will damage their scientific position.

Let me come back now to these questions here. I think these two presentations were marked with the absence of dealing with the issues again, or with the subject. Now, Dr. Dull is extraordinary. On the one hand we are presented with the fact that there is no systematic reporting of laboratory-acquired infections. Not only isn't there systematic reporting, there isn't systematic diagnosis. It is anecdotal.

I did not understand how it is possible to conclude in the absence of any idea of what percentage of the actual cases you are seeing that the incidence of laboratory-acquired infections is going down, when one has no idea about what the incidence is and the efficiency of reporting.

I have had government grants for the last fifteen years, and I have worked with *Shigella,* and *E. coli,* and *Salmonella typhimurium,* and I have never received a communication from the CDC or the NIH asking me, requiring me, hinting that I report cases of laboratory-acquired infections to them. Neither have I received any other communications. I know for a fact that there are many cases of laboratory-acquired infections where I work, and they are not reported at all; they are just known anecdotally. Furthermore, there is a much larger number of infections that are not diagnosed as laboratory-acquired. Somebody works in the lab, they get sick, right? They get a urinary tract infection, they go to the doctor, the doctor gives them Gantrisin. Do we know whether or not that was a laboratory-acquired infection? No.

My wife has a vicious case of bacterial conjunctivitis that is antibiotic-resistant. Does the medical department ever check to see whether that was acquired through me from a microbiological laboratory? No. I mean, it is part of the underdevelopment of occupational medicine in the United States that there is no monitoring or diagnosis of laboratory-acquired infections. In fact, the data on that is very, very weak.

Now, as to Dr. Davis' presentation on the so-called facts on infectious disease. Let me refer to data on, say, three points. One is the question of the exchange of naked DNA in the gut. He mentioned pneumococcus. You will find in the *Handbook of Microbiology, 1974,* an article by Ravin on what is known about the take-up of DNA by bacteria. People have studied this for a long time. It is called transformation. There are no recorded cases, even though there has been a great deal of research, where one species of bacteria such as pneumococcus or hemophilus or something like that takes up DNA from an unrelated species, none at all. No one has ever found it or reported it.

Now, if you take DNA from *E. coli* and ask does it have genes, human genes from DNA, the answer is again no. This is an easy electron microscope hybridization experiment. No one has ever claimed that the best studied *E. coli* have genes in them that are homologous to human DNA. Furthermore, if we were to talk to a geneticist, a microbial geneticist, and ask do we expect that, the answer is no. We know these bacteria make enzymes called restriction enzymes that are used in this research. What do these enzymes do? They chop up foreign DNA. Furthermore, we know experimentally that the DNA is not homologous and that the normal modes of genetic recombination do not operate here. There is no reason to believe that *E. coli* have been routinely taking up and integrating DNA from the human gut.

Now, as to the question of an epidemic, none of us that I know of who have been concerned about the dangers have talked about epidemics. It is the people on the other side who are talking about epidemics. The reason for that is that we have been concerned by those people who are alredy sick from *E. coli* infections. If you go to the *Journal of Infectious Diseases,* you will find that, for example, in Boston one in one hundred hospital admissions suffers from acute enteric infection, and about 25 percent of those are *E. coli* infections. It has been estimated that roughly 30,000 people a year in the United States die right now, each year, from acute *E. coli* infections. These aren't theoretical infections; these are real people who are suffering from real *E. coli* infections.

The fact that these have not spread as an epidemic does not mean they are not a problem.  If you have an acute urinary infection the fact that you may have gotten it through routes other than direct contact from somebody else, from your own flora, doesn't alleviate the fact that you are suffering from the acute infection.  The whole history of antibiotic resistance, the acute increase of *E. coli* as a hospital-acquired infection in the last twenty years, has been shown. It is one of the triumphs of molecular genetics that we know that that is due to the introduction into those cells of a few genes for antibiotic resistance carried by plasmids.  If you cure those cells of those plasmid genes, they are not resistant to antibiotics, they don't make you sick.

I mean part of the basic data of molecular genetics and all of this research around plasmids is that as a matter of fact a very few genes can make a very big difference.  Toxin-plus and you get diptheria, toxin-minus and you don't get diptheria.  Antibiotic-resistant and you are in trouble, antibody-sensitive and you are not.

In the *Journal of Infectious Diseases* there was a recent article on this particular strain of *E. coli* that causes urinary tract infections in women.  So the investigators isolated *E. coli* strains from 149 consecutive women who had suffered urinary tract infections, and they said are they invasive, do they make a toxin?  That is, do they have the properties that have been proposed for infectivity? And the answer was no.  The answer was we don't know exactly why those women are suffering from those urinary tract infections, and it would not necessarily take a very big change in that *E. coli* so that it causes male urinary tract infections.  I wonder how many of the panel would be willing to tolerate a nonepidemic example of that.

RICH:  Dr. Dull, would you like to respond?

BRUCE DULL:  Dr. King is quite right, and he heard me correctly when I said that there are obvious biases in the data.  These arise partly in that our information comes from a voluntary system of reporting of laboratory infections, just as our data on the occurrence of diseases generally are based on a voluntary system of notification. We estimate with respect to the reporting of diseases in general that probably only about 10 percent of illnesses seen by physicians and in hospitals are reported.  However, epidemiologists who deal regularly with fractional data like these may assume, if the data sources are comparable year to year, that important conclusions can be drawn.  Even with the obvious, as well as the subtle, shortcomings in the data, we believe there is enough internal consistency to permit valid conclusions.  For example, in terms of the decline in bacterial infections associated with laboratory practise, a tenfold decline may be numerically questioned.  Whether it is eightfold or twelvefold  in any given year may be less important than if the trend has been consistent in preceding years.  But Dr. King is basically correct in saying that our data are incomplete.  We do rely in this country on voluntary reporting, which has served us well but which, of course, should be improved as much as possible.

DAVIS:  When Dr. King says that there is no uptake of foreign DNA by bacteria in nature, or no routine uptake, it seems to me that is irrelevant to the probability that I was discussing.  When I suggest that it is hard to believe that *E. coli* in nature has never taken up DNA from the mammalian cells that surrounded it, this does not mean that every tenth *E. coli* cell is going to be found to have some mammalian DNA in it.  It means that one out of $10^{15}$ or $10^{20}$ of cells might.  If that gave the cell a selective Darwinian advantage, that would now begin to be found as a predominant strain, but it is not.

So what we are postulating--and the only reason to postulate it is simply the argument that there is fantastic novelty to what we are now able to do in the lab--is an event that is likely to be, from what we know about the properties of the organisms in the gut, namely their low competence for taking up DNA, likely to be a very rare event.

Now that doesn't mean that there should not be experiments done to test more exhaustively for it, and I believe there are such experiments going on, for the uptake of foreign DNA by *E. coli*.  But a negative result would not in any sense controvert the prediction that if *E. coli* is able, under the conditions Stanley Cohen worked out, to take up DNA quite effectively, when you put some calcium ion in the neighborhood, for example, that somewhere in somebody's gut once in a while there is going to be an *E. coli* cell that will take up some DNA.

Now, to move on to the question of epidemics, I don't know about this figure of 30,000 *E. coli* deaths.  Is that something that rings a bell with you?  It sounds a little high to me.

KING:  Well, since they don't monitor that all across the country, it has been generalized from major areas of population.

COHEN:  He said it was an extrapolation from a few cases to the population of the country.

DAVIS:  May I comment on some other aspects, and Dr. Cohen, maybe you can comment on this, because you are working with infectious disease, and I no longer am.  But the point I would like to make is that when Dr. King talks about all these people getting *E. coli* infections, say, in their urinary tracts, and that that somehow has a bearing on recombinant DNA research, I don't see the relevance at all.  We do not have epidemics of urinary tract infection.  Urinary tract infection occurs in people, primarily in females, as a result of usually abnormal conditions that result in some retention of urine.  Therefore normal *E. coli* from the gut get into the urinary tract, are not flushed out fast enough, and maintain a certain level of multiplication in the urinary tract.  I don't know of any evidence that it is special strains of *E. coli* that cause urinary tract infections. So it seems to me the question of what kind of *E. coli* is going to cause urinary tract infection has nothing to do with recombinant DNA.

## HEALTH HAZARDS TO LABOR

Anthony Mazzocchi

*Director, Citizenship-Legislative Department, Oil, Chemical and Atomic Workers International Union*

I am not here to talk about health hazards to labor; I am here to find out whether health hazards really do exist. My statement is one labor union official's response to a scientific debate. First, I would like to commend those responsible for inviting for the first time a representative of possibly potential victims to join in the discussion before the fact--rather than after the fact. Also, we view the debate between the proponents and opponents of this research within the context of our own experiences. As you all argue over each other's credentials, and rather politely, we view that also with great interest, based on our actual experience in the real world of work, where ultimately these processes are carried out.

I don't have any scientific credentials. Therefore, I will speak only to the area in which I think I am qualified, although I have heard a great deal of discussion here from members of the scientific community dealing with areas in which they have little credentials and expressing gross naivete about how part of the real world works.

Our experience in the labor movement with questions of occupational epidemics from cancer and other diseases has been that what was known by scientific investigators was rarely, if ever, conveyed to the population at risk. What we know about practically every epidemic today, and every assault, is based on our own empirical evidence rather than by virtue of scientific investigators allowing us to share in what they knew based on animal experimentation and observation of what happened to humans. So we are very skeptical about any debate about the introduction of a possible potential hazard. It is within this context that we view this debate, because we have been victims, and we have not been part of the decision making in any shape or form to date.

I heard one of the previous speakers use an expression, "if they escape the laboratory." Now, that is a naive assumption I would like to address myself to based on our experience. One can't be serious if one talks about federal regulations, voluntary guidelines as being forms of containment. They have not worked up to date. Based on our actual experience, working people look and laugh at any suggestion that a law or a voluntary guideline could contain a possible contaminant, based

142

on our actual experience.  The Occupational Safety and Health Act was
passed in response to massive indignation of the American work force
over the fact that standards were arrived at through consensus by industry, and there were a great many voluntary guidelines, and there
was voluntary enforcement that did not protect and failed to meet the
test of providing a work place free from hazards.

We look to law as a forum in which we can express our concerns, but
if the population is to expect that the law or voluntary guidelines
will protect them, they are in for a sad realization.

Now, dealing with the pharmaceutical industry who probably will be
involved in this particular research, their track record is poor also.
So if there is a problem with this recombinant DNA, we feel less than
assured that there won't be problems to the work force initially and to
the population in general.  It hasn't worked up to date; there is
nothing to suggest that it will work in this case.

Our reaction to this debate is that when two sections of the scientific community argue over, number one, each other's credibility or
credentials, or whether something is safe or unsafe, I think our posture
has to be that we ought to be very prudent before we proceed.  In our
experience, whenever a new process was introduced that had a possible
danger attached to it, that danger expressed itself in the work place
and in the community by attacking workers and attacking workers'
families.

We do not want to talk about benefits and risks at this point.  We
think that is the second step of a discussion.  We think the first step
of a discussion about recombinant DNA must be whether it is safe or
whether there is a possibility of it being unsafe, and how unsafe it may
be.  Then we ought to discuss societal implications, rather, would we
prefer to do something else about what is affecting us.  We suffer,
as workers, from a cancer epidemic.  We think a better approach to
this problem would be containment in a work place.  We have the technology, we have the methodology; however, there are political and economic
restraints.  We understand that most occupational health and medical
questions are really not scientific questions, and really shouldn't even
be debated by the scientific community, simply because they are economic
and political.

The scientific community has conducted a great many investigations
and, as I said, contained the discussion.  The science community was
aware that cancer in rats took place by virtue of exposure to vinyl
chloride years before we in the work place became aware of it.  We became aware of it through the body-in-the-morgue method, a rather primitive method of finding out the truth.  Each time an epidemic has come
along in the work place, we have had to base our actions on our own
observations.  No one in the community, the scientific community, talked
out of school.  This is the fear we carry into recombinant DNA.  We
have read both sides of the argument, and we see the fact that there is
a great divergence of opinion.  I repeat, our position is to be prudent,
because if this research takes place in industry and there are problems
with it, no set of rules is going to contain it.  Our experience in

the pharmaceutical industry indicates to us that if a contaminant exists, it will show up. What we have to consider is how do we deal with the contaminant. The question raised by a previous speaker as "if they escape" must be posed as "when they escape."

About the discussion over laboratory infections: if there is anything that frightens us, again, it is any statistical information that comes out from the federal government, because most of it has very little basis in fact because those who have the responsibility to report it don't report it. If you think there is an accurate reporting system about what happens at work, that is incredible naivete. There are tens of thousands of victims that have been hidden.

Let me cite one situation. We have 450 workers dying of cancer out of a population of 900. This is work-induced cancer, and this is a situation where federal investigators were aware of the situation; corporate medical investigators were aware of the situation. Discussion went on between the federal government and the corporate executives and the corporate medical executives about the nature of the problem, and yet the people who were at risk were never consulted. Only a young doctor, by virtue of a moral imperative, found the information, violated a federal law, because he was a commissioned officer, and sent it to us. This was the only way we became aware of this epidemic. In fact, these 450 victims would be buried, as they died, without us understanding the relationship between what was going on in that work place and the fact that they ultimately became diseased.

Now, the failure to report is a condition; it is a real condition in the world of work. If an industry is doing something that has an implication for either the workers or the population, they are not going to report it. The question of class actions and civil suits is a real one. The fact of unduly alarming--as industry likes to say--the population or the work force is a real one. So victims aren't reported. We have very little faith in reported diseases. The fact that the Fort Detrick experience was stated as 425 infections, who says so? Who said there were only 425 infections? Who said the community was not infected? Who counts?

If you look at the tools that exist in this country for counting victims, you will see that they are very shoddy and primitive. How many tumor registries are in the United States, real tumor registries, beside the state of Connecticut? The third-party payer system purposely obfuscates data. You can't draw on data from third-party payers. It is not even accumulated. We, who have to deal with this condition, as I say, deal with this whole question out of our experience, and our experience has been poor because those who have known what goes on haven't bothered to tell those of us who are at the point of risk.

Now you are talking about something that goes far beyond the work place. If the critics are correct and the Andromeda scenario has even the merest possibility of occurring, or a variation of it, we will have to assume it will occur, based on our experience.

Our position must be that this debate should be held in the open. It should incorporate more than members of the scientific community, and it shouldn't be just a mere nod to the people who work, by virtue of slotting in one individual on a program such as this. Most of you are not among those who know how that real world of work operates. Those who know are the people out there who work in these factories, the very plants who are sponsors of this particular conference.

Our suggestion is that many of these groups would have to get their own house in order before we would trust them with recombinant DNA research if there is even the slightest possibility of any validity to the observations of the critics of DNA recombinant research. The stakes are rather high if there is even a plausible factor involved in them being right, again based on our own experience.

I am not here to recite a litany of woe of what has happened to working people. The evidence is there. The President of the United States, in grossly understated figures, talks about 100,000 of us who die each year from occupationally induced disease. We say it is low because not everyone is counted. And certainly 100,000 victims a year is not an extrapolated figure. I am talking about the President's report, and we have serious quarrel with whether they have counted all of the victims, and since less than 5 percent of occupational Workmen's Compensation cases are from occupationally reported disease, we would think that many of these cases don't find their way into the system.

Certainly at this point we cannot enter into the scientific question here--I read a little about recombinant DNA research. I had a great deal of difficulty even pronouncing some of the words. However, based on our experience, I would end with this observation. I have lived long enough to remember the birth of the atomic industry, and we were given all sorts of reassurances about what the safety implications were involved there. I sat in scientific debates such as this as a member of an audience twenty years ago, and have lived long enough to see that my worst fears have been borne out in certain aspects of that particular industry. I have certainly seen it in the chemical industry, where we didn't even have the benefit of a debate over the introduction of substances.

Our view is that the burden of proof lies with those who wish to introduce a new process if there is a possible hazard to it. That view is based on our hard experience, our very hard experience, looking back over that long toll of victims. We think the debate should go on, we think it should be expanded. We are not reassured by what we read and hear at this point. I listened to some of the comments this morning, and I heard reference again to regulation. I heard them talking about congressional subcommittees. I would say again--that doesn't reassure us. We think you are being naive if you think really that regulations, whether they are voluntary or mandatory, will contain something that may be hazardous to the population.

Let us continue the debate. Let us expand the debate, but certainly, let us be prudent and not proceed until we have resolved all the outstanding questions beyond the shadow of a doubt. Our experience with cancer has been that the results are irreversible. And until such time as the working population--and it is a large population--can be assured there are not risks to ourselves and our families, continue the debate.


# AN ENVIRONMENTAL OVERVIEW
# OF RESEARCH WITH RECOMBINANT DNA

---

## Delbert Barth

*Deputy Assistant Administrator for Research and Development, Environmental Protection Agency*


In addition to my other duties, I have been designated as EPA's official representative to the Interagency Committee on Recombinant DNA Research, which is chaired by Dr. Fredrickson, Director, NIH. A major portion of my presentation today will be concerned with the activities of the Interagency Committee. It must be emphasized, however, that my remarks represent EPA's views and opinions and may not, in all instances, reflect consensus positions of the Interagency Committee.

EPA recognizes that research with recombinant DNA is under way and is likely to continue and be expanded, both in the United States and abroad. It is our view that this research has a sufficient amount of risk associated with it that some regulatory controls, over and above those now in force, should be instituted. During the course of this brief presentation, I shall outline for you some of our views with respect to this important matter.


## BACKGROUND

The NIH guidelines for recombinant DNA and the draft Environmental Impact Statement (EIS) do list the hazards of this type of research; however, most of this information is entirely speculative. Of major concern is the potential release into the environment of toxic or pathogenic agents resulting from such recombinant DNA experiments or technology.

When reviewing the EIS, EPA expressed the following concerns and reservations:

● Insufficient knowledge seems to exist now to assure that normally harmless strains of *E. coli* or other host organisms will not be transformed into virulent pathogens that could find a niche for multiplication if accidentally released into the environment. Therefore, we recommend that very high priority be given to search for the knowledge or evidence needed to demonstrate safety of the biological containment concept.

● The guidelines do not yet apply to all laboratories and researchers engaged in recombinant DNA research.

• Even if the guidelines were made universally applicable, it would be very difficult for any regulatory agency to enforce them. Observance of the guidelines depends primarily upon peer pressure without the force of law. Observance also depends on voluntary actions of individuals within laboratories to which the guidelines apply. No truly reliable means seem to exist for detecting or giving early warning of accidental release to the environment of potentially hazardous material. Furthermore, no penalties--other than loss of federal financial support--seem to exist to deter individuals who deliberately do not honor the guidelines.

• We are concerned about the apparent uncertainties in the ability to assess the probability or degree of hazards resulting from experiments carried out under less than the most stringent containment conditions.

• We are apprehensive about problems of setting standards for regulation of industrial applications of the new technology expected to flow from recombinant DNA research.

• Neither the EIS nor guidelines adequately discuss environmental-spill contingency plans.


INTERAGENCY COMMITTEE ON RECOMBINANT DNA RESEARCH

A memorandum signed by President Ford to the heads of departments and agencies, dated September 22, 1976, states:

On July 23, 1976, NIH released guidelines for the conduct of research on recombinant DNA experiments. These guidelines establish carefully controlled conditions for experiments in which foreign genes are inserted into microorganisms, such as bacteria. The objective of the guidelines is the containment of these possibly dangerous organisms while permitting research of great potential benefit to mankind.

It is recognized in this letter that there are potential risks for these techniques as well as potential benefits.

The Department of Health, Education, and Welfare was given the lead for setting up an interagency committee to review federal policies for conducting research in recombinant DNA. President Ford urged all other departments to cooperate fully with HEW.

On November 4, 1976, the first meeting of the Interagency Committee on Recombinant DNA Research was held. A summary of that meeting follows.

The charge of the Committee was to review the nature and scope of recombinant DNA research in both the public and private sectors; to determine the applicability of the NIH guidelines to govern such research; and to recommend legislative or executive action needed to ensure compliance with the standards as set.

The Chairman, Dr. Fredrickson, pointed out the need to consider the following list of functions and processes required in the regulation of recombinant DNA:

- Registration of activity
- Certification of containment standards
- Oversight of investigators and institutions
- Formulation of an appellate mechanism to the above
- Requirements for safety education and training
- Development of safer hosts and vectors
- Establishment of a mechanism to provide hosts and vectors
- Exchange of information
- Establishment of international liaison
- Extension of the guidelines throughout the public and private sector
- Placement of ultimate authority

The NIH General Counsel reviewed potential regulatory authorities of the following governmental agencies:

- Center for Disease Control
- Food and Drug Administration
- Department of Transportation
- Environmental Protection Agency
- Department of Agriculture
- Occupational Safety and Health Administration

Discussions followed at the meeting with regard to the need to extend the guidelines, or some comparable guide and regulatory action, to the private sector.

At the end of the meeting, Dr. Fredrickson requested the following:

- Written statement analyzing the nature and extent of actual or planned recombinant DNA research and the role of each agency vis-à-vis the eleven functions/processes previously mentioned.
- Regulatory agencies to submit a written statement analyzing the authority and role of the agency in the regulation of recombinant DNA research.

Many activities have already been compiled. EPA is presently conducting no research in this area but we do have applicable regulatory authority for parts or perhaps nearly all of it.

It is expected that the final Committee report will be out by mid to late March.


LEGAL AUTHORITIES OF EPA POSSIBLY APPLICABLE

At this time, we know very little about the real hazards associated with recombinant DNA. Under Section 112 of the Clean Air Act, the administrator may list hazardous air pollutants and set emission standards for such pollutants. However, before standards could be promulgated, the Agency would need to know a great deal more than it does now about the hazards associated with recombinant DNA and the manufacturing process itself. Likewise, under Section 307 of the Federal Water Pollution Control Act, the administrator may set effluent standards for toxic pollutants. The same applies here as with the section under the Clean Air Act that the Agency needs to be better informed about the hazards of recombinant DNA before it could promulgate such standards. It would be very difficult to monitor effluent or discharge from the laboratories.

The passage of the Toxic Substances Control Act made available a broad range of regulatory options with respect to laboratory experimentation and practical application of recombinant DNA technologies. Under Section 6 of TSCA, if the administrator finds that there is a reasonable basis to conclude that the "manufacture, distribution in commerce, use, or disposal" of a "chemical substance" will present an unreasonable risk of injury to health or the environment, the administrator shall by rule apply the least burdensome of eight specified requirements. Section 3 of TSCA defines the term "chemical substance" as any organic or inorganic substance of a particular molecular identity.

Provisions of Section 6 of TSCA apply only if the "manufacture, processing, distribution in commerce, use, or disposal of a chemical substance will present an unreasonable risk of injury to health or the environment."

Section 5 appears to exclude laboratory research with recombinant DNA. Section 5 requires *inter alia* premarket notice or in this case preexperiment notice.

Congress intended Sections 6 and 7 of TSCA to apply to laboratory research.

TSCA could regulate use of research in the laboratory by requiring warnings, record keeping, and regulating disposal. EPA might need additional authority to force the researcher to follow the NIH guidelines in particular.

With regard to practical applications of recombinant DNA technologies, the same statutory tools are available to regulate the commercial manufacture, distribution, and use of recombinant DNA as are available to regulate laboratory experiments with recombinant DNA.

Like our interpretation of any of our statutes, this interpretation of TSCA is subject to legal challenge. No single agency felt they had all authority necessary for regulating research on recombinant DNA. It is doubtful that EPA has authority to compile a registry of persons doing recombinant DNA research. One of the major problems appears to be that, without knowing who is doing the research, we cannot promulgate regulations.

New legislation is being sought without being specific as to what that legislation should say. The trend of the subcommittee seems to appear to be to have NIH set the standards and CDC enforce them. Tightening up TSCA might be an answer, and this was the suggestion given by Mr. Zimmerman from the Environmental Defense Fund.

If EPA acts under TSCA to regulate recombinant DNA research, it will have to perform a fine waltz with OSHA. For things that appear to come within OSHA's jurisdiction, EPA would have to refer the matter to OSHA for its determination. Hopefully, we could work out some sort of inter-agency agreement.

## DISCUSSION

RICHARD POLLACK, Fusion Energy Foundation: Despite the fact that this entire session is devoted to hammering out the issues, there has been scheduled for several days now a press conference which is going to take place next Monday. At that press conference Friends of the Earth, National Resources Defense Council, and several other Naderite group-ings are going to call for a moratorium on all recombinant DNA re-search throughout the world.

In light of what has been discussed here by Dr. Davis, by Dr. Dull, and by various other people, we know although there are certain risks associated with recombinant DNA research, it is actually necessary for breakthroughs of a fundamental nature in the biological sciences. The problem of cancer, whether we get rid of mutagens or not, is going to be with humans for a long time, and so on.

The real point about this research is very straightforward. If this research doesn't go on you have a policy of deindustrialization, a lack of commitment to technological progress and the well-being that that brings. If you don't identify the recombinant fight as that fight, you are always fighting a rearguard action. It is that point that I address to this panel. The point about it is with the Fusion Energy Foundation forming a biological sciences sector, which calls on industry, calls on science, and calls on labor to advance the fight for progress; it is in that context that the recombinant fight can be won.

STANLEY COHEN: It is my understanding from the information sent to the Panel for Inquiry prior to the meeting that the Panel would be respon-sible for attempting to hold speakers and others to a high standard of documentation for statements. Therefore I would like to comment on two statements that were made in the last discussion. One is the point made by Professor Wald: that it makes little difference what the reality of the situation is regarding the intimidation of young faculty; what really counts is what appears to be the case. I would argue that the facts are indeed more important than the ap-pearance, and I would urge Professor Wald to document his assertions, which really make very serious charges by innuendo. I think it important that we identify what is fact and what is opinion in these discussions.

Secondly, I wanted to comment on Jonathan King's statement that there is no evidence that bacterial cells can accept DNA by transformation from unrelated species. That statement simply is not factually correct, and this evening there is a workshop at which information documenting that will be presented.

And then finally, a comment on the statement by Mr. Mazzocchi. I attended the workshop chairmen's panel and I regret that I missed the beginning of your talk, but I did return in time to hear the end, which included your statement that "the burden of proof is on those who want to introduce a new process. Let us not proceed until we have resolved all the outstanding questions beyond the shadow of a doubt."

I share your concerns about the protection of workers and others, but at the same time I feel that the questions being raised will not and cannot be answered beyond the shadow of a doubt. I would guess there are few questions involving worker protection in any other area that can be answered with such certainty. I would like to know from you, since you did make the statement, what kinds of information you believe are needed to resolve this question, or other questions of safety.

MAZZOCCHI: I think that once workers have a chance to look at all the evidence by both parties, that the resolution of beyond the shadow of a doubt will be made by the population through its mass political forum. That is how it is going to be resolved. What we object to is the resolution of these questions by a small body of people holding discussions without including us in. I can't define "beyond the shadow of a doubt." The people of the country, at some point in time, when they are in possession of all the facts, will register whether we should proceed or should not proceed if truly the debate takes place out in the open, and if truly the decisions are held up until such time as the population has a chance to have their own will carried out through the forums that exist. There is a political process, but I would like to remind you that up to now this process has not occurred. We have responded after the fact in every single instance, and that is the type of arrogance that we are tired of. We are tired of that body-in-the-morgue method, which I would suggest is a truly unscientific approach.

I would like to make this comment. The scientific community has treated us with less than candor about these questions, and that is the experience that we bring to this meeting. You have not been candid--and I use you generally, certainly not implicating everyone-- but scientific investigators, up to date for the overwhelming most part, have not discussed with victims the fact of their experiments, what they have found and their implications. And it is that experience we carry into this meeting. And "beyond the shadow of a doubt" has to be viewed within this context.

COHEN: One last comment on scientific candor in this issue. Dr. Singer noted last night that this issue appears to be different than some of the other issues that you have been talking about.

MAZZOCCHI:  I see no difference at all.  If this issue--

COHEN:  The first concerns about this issue were raised by scientists, unlike some of the other issues that you have cited.  That seems to me to be a very substantive difference.

MAZZOCCHI:  You missed the first part of my talk.  I made an expression of appreciation over the fact that for the first time we were included in on the discussion before the fact rather than after the fact.

KURT MISLOW, Princeton University:  As I have examined this meeting and I have listened to everyone, it is clear that the orthogonal parties are in fact what we might call technological optimists and technological pessimists.  You know, the optimist is someone who thinks this is the best of all possible worlds, and a pessimist is one who thinks that this might be true.

So I myself actually happen to be a pessimist.  I have to be anecdotal here.  When I was a kid my father showed me a pencil and he said, "Son, the energy that is in this pencil can drive a steamship across the ocean and back."  I was so impressed that I decided to become a scientist.  This pencil has become a sword of Damocles over my children.

I recognize the power of technology, and I recognize the power of science.  I think that the benefits and risks that should be discussed, in fact, are the benefits and risks of application.  I consider a risk of application to be as real as the benefit of application.

In any event, the point that I am driving toward is that we ought to be more than cautious.  We ought to be more than conservative.  It is quite right that all of us ought to be involved in this process of deliberation.  The particular point at issue that I want to address is one concerning the NIH guidelines, because this is what has been discussed in this particular section of the meeting.

There is sort of an aura of sacrosanctity about these guidelines which I find somewhat overwhelming, and would probably accept if I didn't know any better.  As I say, I am not a biochemist, but I have watched the proceedings.  I have in my hand here a document entitled "Recommendations for the Conduct of Research with Biohazardous Materials at Princeton University."  This report, which is conceded by a number of outsiders to be an unusually responsible and carefully constructed report, begins by stating that these techniques will have an influence on the biological sciences at least as profound as the discovery of the basic structure and function of DNA.  It goes on to say that they hold a key not only to the central questions in molecular genetics, but the answer to wider processes regulating development, growth, and form.

"Given that possible hazards exist, safety is likely to lie in the acquisition of a better understanding at issue."  And then in its recommendations it states, in part:

The committee recommends that research proposals for work on *in vitro* recombinant DNA at Princeton University be reviewed, one by one, and levels of physical and biological containment specified. It is recommended that a minimum set of requirements (a "floor") be set, and that this floor be more conservative than the NIH guidelines in two significant respects: (1) In certain situations where the NIH guidelines allow one additional level of physical containment to be exchanged for one additional level of biological containment, only the higher level of biological containment will be allowed. (The committee believes biological containment is more effective and more predictable than physical containment.) (2) In the cases where the NIH guidelines draw distinctions between nonembryonic and embryonic or germline tissue, the standards appropriate to nonembryonic tissue always apply. In making this recommendation, the committee is aware that the standards of physical containment in a P3 facility at Princeton University are also recommended to be more strict than the minimum P3 defined by the NIH guidelines.

I won't go into details that are contained in the report. My major point is this. There is agreement as to what constitutes a safe set of guidelines, and I argue that this report provides evidence that there can be more than one set of guidelines considered safe. Therefore our goal should be to explore this question of guidelines in much greater detail, and of course keeping in mind as well the other issues that have been raised at this conference.

BARTH: I am not sure that my presentation was being referred to, but all of the discussions which we have held within EPA and also within the the NIH Interagency Committee recognizes that something like the guidelines must be used. All kinds of legislation which we are presently examining calls for allowing the possibility of amendments to these guidelines for cause. So any of these comments that are made have to to evaluated by various people, and there will be a mechanism, I believe, in whatever law finally is passed, to allow for fine tuning the guidelines in any areas where it is not considered to be safe enough as presently written.

DAVIS: Well, as Dr. Mislow said, there is room for more than one kind of guidelines. Obviously in the face of the amount of concern that there is in the public, as I mentioned, I am not being realistic in suggesting the possibility in the near future of moving toward less stringent guidelines. I suggested that as time passes we will have more experience and they will become less stringent.
    I just hope that what will not happen will be that something like the Cambridge City Council's committee, which did a magnificent job, comes through and says by and large the scientists have been responsible. They have come out with a good set of guidelines, but we don't want to just rubber stamp them; we want to make them a

little bit tighter.  Now there is something going on at some state
levels, and they want to make them a little tighter, and the federal
government may want to make them a bit tighter, and the federal
being virtuous and protecting their constituency, the maxim being that
you cannot be too careful.  But you can be too careful, because the
more careful you are the more expensive it is, and if you get very,
very careful it is paralytic.

CHRIS OLIVER, Doctor of Internal Medicine, Massachusetts General Hospital:
As a practicing primary-care physician, I have a few concerns that
I would like to express.  One of these concerns is the fact that as
I have been sitting through these meetings since last evening, the only
clearly predictable benefit of recombinant DNA research that I have
heard stated is "knowledge of the human genome."  I have very real
questions about the practical benefit of this knowledge to the 3,000
or 4,000 patients that I see each year.

I do have some questions as to what degree this knowledge is self-
serving, both in terms of activity of the scientists involved, and also
in terms of profit motive.

I have a second concern, and that concern is the effect of the
"enfeebled K-12 *E. coli* strain" on the large numbers of enfeebled
hosts that I see every day.  These enfeebled hosts include people with
diabetes, people with rheumatoid arthritis, people with systemic lupus,
with acute glomerulonephritis, with cancer, who are receiving chemotherapy, and the elderly, just to include a few.

I also feel that given the fact that one of the proposed benefits
of this research is medical cure and improvement in the health care
system of this country, more practicing physicians should be included
in the decisions that are currently being made.  As I review the
list of people who were involved in planning this Forum, I see very
few people who seem to be practicing primary-care medicine.  I feel
that input from these people would be most beneficial.

MICHAEL MINK, student in microbiology, University of Michigan:  I have
a question for Dr. Davis, which dates back to using K-12 for these
experiments because K-12 is unable to habitate the gut.  Many bacteria
have the ability if they have a plasmid to donate the plasmid to a
recipient bacteria.  If there is a laboratory infection of a K-12
bacteria containing a plasmid, could it donate it to another *E. coli*
habitating the gut?  And if it can, that means K-12 would not be a
good bacteria to use in these experiments.

DAVIS:  That is a point that I think I covered too briefly in my talk.
I wouldn't draw your conclusion that K-12 wouldn't be any good if this
can happen, because you are, I believe, putting in qualitative terms
what is fundamentally a quantitative question.  The question is:
With how high a frequency can K-12 donate its plasmid to another strain?
And I inserted the statement in my talk that the most important question is not the survival of K-12, it is the frequency with which K-12

will give its plasmid to other strains. I would like to see more work done on it. But the work that has been done so far shows that under conditions where the EK2 strain of K-12 is not given the very special nutrients that it needs in the laboratory and that it will not find in the gut; under those conditions it not only bursts rather quickly, but it also is not able to donate any normal plasmid in it with any great efficiency.

In addition, the kinds of plasmids that are being used are plasmids that do not code. They have been deprived of their initial ability to code for their own transmission. So in order for an EK2 K-12 strain with recombinant DNA in its EK2 plasmid to cause transfer of those genes to a normal, well-adapted *E. coli,* which is what you are worrying about and what I would worry about, you have a multiplicity of very low probabilities. I am not going to try to put numbers on them, but they would be extremely low indeed.

MINK: I have one more point, that although these plasmids may be non-transmissible, if it is in a host with a transmissible plasmid it can be cotransduced, which is some work that is being done at the University of Michigan right now. A transmissible plasmid can donate its transmitting ability to nontransmissible plasmids in the host.

DAVIS: Again you are putting in qualitative terms a quantitative question. Any bacterial cell can in principle take up a complete, nondefective plasmid from some other cell, which can in turn pick up in principle any gene, which can then transfer it to some other organism. It is a question of rate. If you are putting *E. coli* EK2 K-12 strains into a human gut, and they survive for a very short time, then during that time the chances of their surviving long enough to pick up a good plasmid with a very low frequency, and then with another low frequency to transfer that plasmid as dying cells to some other cell, the combination of those is exceedingly low.

ROY CURTISS, University of Alabama Medical Center: First, I think it is important to stipulate that both in the U.S. and the British guidelines it is not permissible to clone DNA into a bacterium that already has a transmissible plasmid. Consequently, in using what we call a nonconjugative cloning vector, which cannot be transferred by itself, a cell having that first has to acquire a conjugative plasmid from some wild-type enteric microorganism encountered in the natural environment if the microbe you are using for the cloning experiments escapes the lab. And then if that happens it must transmit the DNA in this plasmid vector to some other host cell that can survive.

We have evaluated the frequencies with which this occurs under a variety of conditions. I think Dr. Davis has paraphrased what we have done very well. But for a disabled strain such as $\chi$ 1776 we have estimated that the probability that such a nonconjugative plasmid cloning vector would be transmitted out is something between $10^{-16}$ to $10^{-20}$ per surviving bacterium per day. The point that he has made is that these cells can't even survive for a day, and so rapid death in conjunction with these low probabilities makes this something that is unlikely to occur anywhere on the face of the earth with any measurable frequency, certainly.

JEFF JOHNSON: I am just a citizen, and I don't know if I am really qualified to be asking some of you these questions. I would like to know through the knowledge of your experimentations, what would be the probability of this experimentation happening somewhere else in the world? I know it is happening in other places in the world, and I want to know what would keep it from spreading if it is under-taken in other areas of the world, you see, because Hong Kong flu was never isolated to Hong Kong.

And, as a little food for thought, if we understand the patterns throughout the history of mankind, we have to understand the degenera-tive brain of man; and understanding the degenerative brain of man, we have to be wary of further experimentation. Also, if we ignore these patterns found throughout history, on which we have built the country and everything else, then I suggest that the fears of mankind will be rid through experimentation on DNA, because the universe shall be rid of mankind as we know it today.

## POTENTIAL BENEFITS

Irving S. Johnson

*Vice President of Research, Lilly Research Laboratories, Eli Lilly and Company*

To interject the religious theme which some participants felt was lacking in earlier sessions of this Forum, I feel a little bit like a Christian walking not into the Forum, but the Colosseum. I have been asked to discuss the possible benefits of insulin production by a microbial organism, but I think first we have to assess whether this need exists. I would like to begin by making a statement to the diabetic population of the United States, who probably are not represented in this Forum, were not represented at Asilomar, and neither, might I say, were we. There is no shortage of insulin at the present time. There was no shortage of energy a few years ago. Our concern is whether there will be a time when a shortage of insulin may exist. I would like to put the prevalence of diabetes, particularly diabetes mellitus, and the availability of glandular insulin into some perspective.

It is difficult to get hard figures, particularly worldwide figures, for a nonreportable disease. I can't say that the data in Table 1 are completely accurate, but they are our best estimates based upon information from a number of sources.[1,2]

The worldwide incidence of diabetes is something on the order of 60 million. About 35 million of these people live in the so-called less-developed countries. I believe that figure will increase significantly

TABLE 1  Estimated Incidence of Diabetes

| | |
|---|---:|
| World total | 60,000,000 |
|   Less-developed countries | 35,000,000 |
|   Developed countries | 25,000,000 |
|     Undiagnosed | 10,000,000 |
|     Diagnosed | 15,000,000 |
|       Insulin-treated | 5,000,000 |
|   United States | 5,000,000 |
|     Undiagnosed | 1,500,000 |
|     Diagnosed | 3,500,000 |
|       Insulin-treated | 1,250,000 |

when their diet becomes more "western," with a higher fat and carbo-hydrate content.

The figures for the United States are more reliable, and we feel much more confident about their accuracy. There are at least 5 million diabetics in the United States. In 1976 over a quarter-million of them received insulin, including about 100,000 who were children.

There is no alternative therapy for the insulin-requiring diabetic. In milder forms of diabetes there are oral hypoglycemic agents, which are declining in their applicability, as well as a rigid diet. But there is no replacement, no substitute at the present time for the hormone itself. This fact could change with future research, as I will discuss shortly.

The data in Figure 17 may not be completely accurate either, but again they are the best that we have been able to obtain from a number of sources.[3] All insulin is presently obtained by extraction of pancreas glands, primarily from swine and cattle. Figure 17 shows data for the last seven years, and projections for the next four. The solid-colored areas are based on statistics gathered from a number of different sources. The double-hatched areas are estimates of events that occurred in the past, and the single-hatched areas are our projections for the next four years.

It is obvious to anyone examining this data that if you could get at all of these sources, one could double the amount of available insulin. There are a number of reasons why all of this potential cannot be realized. In the first place, many glands are not collected. They are not collected for a number of reasons: the slaughter does not occur in a USDA-inspected facility; it is condemned because it is contaminated or abnormal in some way; the slaughter occurs in small, local abattoirs where collection is impractical; or because of basic inefficiencies in the packing plant collection process.

The bottom line illustrates the rate at which the insulin requirements are increasing. It is important to note that the worldwide population is increasing at about a 2 percent rate, and the U.S. population at

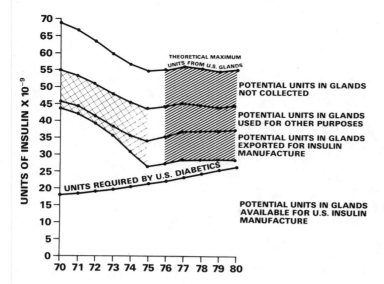

FIGURE 17  Estimated U.S. insulin supply and consumption.

about a 1 percent rate.  In contrast, the diabetic population in the United States is increasing at about a 4 to 5 percent rate, and I see little evidence that this is about to change.

It is also obvious from Figure 17 that some glands are used for other purposes.  Lilly, as well as other insulin-manufacturing concerns, has made and will continue to make serious attempts to divert some of these noninsulin uses of the pancreas glands back into preparation of insulin.  This has not been an easy task.  This is particularly true in terms of Japanese usage.  They like to produce pancreatin, kallikrein, and a number of other enzymes from pancreas glands, and there are some populations that use the gland for food.  You will also note that some of the available glands in the United States are exported to foreign manufacturers who do convert them to insulin.

During the 1970s there were a number of events that had a serious impact on gland supply and quality.  One of these was drought conditions in many parts of the world, which adversely affected grazing areas, resulting in a significant reduction of the cattle population.  Poor growing conditions in grain belt areas of the world resulted in a dramatic reduction of the supply of feed grains normally used in the cattle and swine industries.

The amount of insulin in the pancreas of animals kept on different diets differs significantly.  Graze-fed animals, for example, have a lower insulin content per gland and per gram of protein of the gland than animals that are fed high-carbohydrate and high-fat diets under feedlot conditions.

Another factor that contributed to the overall decrease in insulin yield per pancreas was the introduction in the early 1970s by Lilly and some other manufacturers of a final product of much higher purity.  This has resulted in significant reductions of lipoatrophy and insulin allergy in diabetic patients, but it also results in a reduced insulin yield per gland.

All of these factors have contributed to a fairly striking decrease in the theoretical insulin potential in the first half of this decade. We feel that these factors have probably stabilized, and are projecting a fairly stable situation for the remainder of the decade. Our concern is that many of the factors controlling gland availability are unpredictable and uncontrollable by man. I have mentioned drought and severe weather conditions. We cannot predict with great certainty whether farmers will continue to raise cattle, if in fact it costs more to raise them than they receive when they sell them. These are uncontrollable and unpredictable factors, which could change in either direction. It may be that we will have great weather, and farmers will produce more and more animals.

This is data from the United States, only, and it is obvious that there are many parts of the world in which large numbers of animals are slaughtered that could contribute potentially to the overall pool of available glands. Western Europe probably collects glands at about the same efficiency that we do, but many of the countries where there is significant potential for glands are countries that do not have modern, centralized slaughterhouses with facilities for rapid-freezing the glands or shipment of the glands in a frozen state. So while this potential exists, I think it would be difficult to realize it quickly.

Based on the uncertainty of these sources, and the possible convergence of the projections for insulin requirement and availability, it seems not only prudent but the responsibility of the scientific community to at least consider the development of other contingencies for maintaining an adequate supply of this essential hormone. We intend to do our part in that regard. I have emphasized bovine and porcine pancreas as the main source, and another obvious question is, aren't there other sources?

At the top of Figure 18 is a schematic representation of the insulin molecule, showing specifically the positions of amino acid residues 8, 9, and 10 in the A chain and residue 30 in the B chain. In most mammalian species all the other 47 positions of this molecule are invariant.[4]

Parenteral administration of foreign proteins has a number of potential hazards, as I think you are all aware. The further you diverge from the structure of the human insulin molecule, the greater your probability of having some medical problem. That is why insulin is one of the most clinically acceptable varieties.

The only other animal, I believe, from which one could conceivably realize appreciable quantities of glands is sheep. There are currently attempts going on in Australia and New Zealand to increase the availability of sheep glands. The limited amount of insulin that can be obtained from the relatively small sheep pancreas would unquestionably result in a more expensive product than porcine or bovine insulin and would probably be a less effective product as well.

160

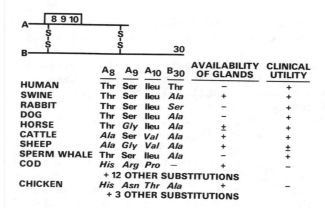

| | $A_8$ | $A_9$ | $A_{10}$ | $B_{30}$ | AVAILABILITY OF GLANDS | CLINICAL UTILITY |
|---|---|---|---|---|---|---|
| HUMAN | Thr | Ser | Ileu | Thr | − | + |
| SWINE | Thr | Ser | Ileu | Ala | + | + |
| RABBIT | Thr | Ser | Ileu | Ser | − | + |
| DOG | Thr | Ser | Ileu | Ala | − | + |
| HORSE | Thr | Gly | Ileu | Ala | ± | + |
| CATTLE | Ala | Ser | Val | Ala | + | + |
| SHEEP | Ala | Gly | Val | Ala | + | ± |
| SPERM WHALE | Thr | Ser | Ileu | Ala | − | + |
| COD | His | Arg | Pro | — | + | − |
| | + 12 OTHER SUBSTITUTIONS | | | | | |
| CHICKEN | His | Asn | Thr | Ala | + | − |
| | + 3 OTHER SUBSTITUTIONS | | | | | |

FIGURE 18   Potential sources of insulin.

The next factor one might consider is: are there any alternatives to other sources of glandular insulin?

I think there could be, and some of them are:

1. Improved utilization of insulin
2. Drugs mimicking insulin
3. Surgically transplanted pancreas
4. Transplanted islets
5. Chemically synthesized insulin
6. Tissue culture production
7. Insulin production by genetically manipulated cells

Many of these possible alternatives are as uncertain as some of the unpredictable factors that affect the availability of glands. There are studies and programs under way that would attempt to improve the utilization of insulin by implanted glucose monitors and pumps, which would put out small amounts of insulin on demand, with the hope that less insulin would be required. There is certainly the possibility of increased utilization by the combination with peptides such as somatostatin--if a longer-acting somatostatin could be made available.

There is a theoretical possibility of drugs mimicking insulin due to reaction with the insulin receptor. Such a reaction could produce a similar sort of biological activity. This is an attractive goal but one which has not as yet led to the development of a drug. A number of laboratories, including ours, are interested in this possibility.

Pancreas glands have been surgically transplanted in patients with physiological evidence of function for short periods of time. These glandular grafts suffer from the rejection phenomenon, and until this problem is solved, which might be the subject of another Forum, pancreatic transplantation is probably not a viable alternative. The same analysis can be made for transplanted islets. Islets transplanted into syngeneic animals function well in laboratory studies. If the rejection phenomenon could be resolved, this is a therapeutic possibility.

Availability of material for transplantation would be a logistic problem in its own right, but perhaps not an impossible one. Similar problems were encountered with kidney transplants several years ago. Another potential problem with transplants, of course, is that there may be factors in the body that will impair the transplant, as they may have in fact impaired the function of the patient's own gland.

Another possibility is the total chemical synthesis of insulin, which has been accomplished in at least three and perhaps four laboratories of which I am aware. It was a monumental laboratory achievement, as was the synthesis of chlorophyll, but it can be achieved only with great effort and at great expense, with miserable yields, and I don't see it as a viable alternative at this point in time.

A number of hormones have been produced by cells in tissue culture, but the least satisfactory in this regard has been, in fact, insulin. The beta cell in culture has not propagated well while maintaining its ability to synthesize and release insulin.

The major theme of this conference offers a different opportunity for manipulating cells to produce materials such as insulin, and it seems to us that in order to ensure an adequate supply of insulin for future generations of diabetics, that we should at least consider recombinant DNA as one of several contingencies.

I pointed out to you the difference in rate of growth of the diabetic population compared to the general population of both the United States and the world. This is compounded, of course, by the fact that we are treating these people. The fact that the juvenile diabetic is treated as effectively as he or she is, the fact that the people who require insulin don't die, survive and have children, must at least be increasing the gene pool for susceptibility to diabetes in the general population. One might question whether that was a good thing to do. If I were a diabetic, I think I would feel that it was. But the question has been raised in the Forum of whether one should in fact treat genetic deficiency diseases. It is possible, as we think about it carefully and systematically, that we could.

It may be a different question whether we should. I am not answering that question, but I would be delighted for some of the members of this audience to consider it. Related to this problem, in Table 2 I have listed a number of human proteins that could theoretically be produced by recombinant technology and in some cases used for the treatment of genetic deficiency diseases. This list is not all-inclusive, and it is not meant to be. It is merely meant to bring out some examples. There are other proteins and hormones worthy of consideration for production by this technology. Human growth hormones can be obtained now only by extraction of pituitary glands from cadavers. Although the number of people who require this hormone is limited, the need from those who do is extremely important.

There are a number of factors which could theoretically--and I emphasize theoretically--be provided by this type of technology, which otherwise will probably never be commercially available.

TABLE 2   Medically Important Candidates for Production by Recombinant Techniques

---

1. Hormones
   A. Growth hormone
   B. Glucagon

2. Coagulation factors
   A. Antihemophilic globulin (factor VIII)
   B. Christmas factor (factor IX)
   C. Plasminogen activator
   D. Plasmin

3. Hereditary disease replacement enzymes
   A. Gal-1-PO$_4$ uridyl transferase
   B. Glucose 6-phosphatase
   C. Phenylalanine hydroxylase

4. Immunological factors
   A. $\gamma$-globulin
   B. Interferon

---

I would like to discuss briefly my view of the biohazards of such an operation. Following is at least a partial list of tests that could be applied to a recombinant organism that might be a potential candidate for use in a process of commercial importance:

1. Nutritional and environmental requirements
2. Gastrointestinal tract colonization
3. Antibiotic sensitivity
4. Animal infectivity, pathogenicity, and toxicity
   A. Normals
   B. Germ-free
   C. Immunodeficient
5. Production and absorption of metabolites from gastrointestinal tract

I believe an organism that might be used for this purpose should be subjected to very severe testing for nutritional and environmental requirements. It is possible to evaluate the ability to colonize the gastrointestinal tract in the laboratory, under controlled conditions. We are able to assay for antibiotic sensitivity, and I might point out that it is routine practice in medical microbiology in evaluating a new antibiotic to determine whether or not it kills organisms that are resistant to older antibiotics. This is one of the merits of adding another antibiotic to the armamentarium.

One can carry out carefully controlled experiments for infectivity, pathogenicity, and toxicity in the laboratory. I have suggested this might be done in normal, germ-free, and immunodeficient animals. If we consider the specific example of insulin, it is possible to measure the production and absorption of such a molecule in the gastrointestinal tract. Very sensitive species-specific radioimmunoassays are available, which, in the unlikely event of significant gastrointestinal tract colonization by an insulin-producing organism, could detect whether any insulin was being produced and whether any was being absorbed.

We have made one calculation concerning this unlikely event, based primarily upon the amount of protein in an *E. coli*, the number of *E. coli* normally in the gastrointestinal tract, and the amount of protein synthesis controlled by the tryp-operon, which is a very significant amount of protein. If you make these assumptions, in a wild-type organism,[5-8] the most insulin that could possibly be present in the human gastrointestinal tract is something in the order of ten units. There have been studies in both animal and man in which hundreds of milligrams, not units, of insulin have been infused directly into the gastrointestinal tract, bypassing the stomach, and levels of ten units or in fact levels of milligrams would not cause a physiological effect, nor would one be detected, based upon this data.

I thought it might also be useful to point out another fact that may not be known by attendees of this Forum. Medically useful agents have already been produced by large-scale fermentation of *E. coli*. One of these agents is L-asparaginase, an enzyme used in the treatment of malignant disease. While we no longer produce this enzyme, we did so at one time. We did it primarily because one of our biochemists, whose eyes light up at the word enzyme, discovered he could crystallize it in a state of unusual purity. The material available at that time for treatment of patients was fairly crude with many side effects. It seemed incumbent upon us to find out if material of greater purity had any therapeutic advantage. We did this, and it turned out it did not.

Table 3 offers some information on the organisms used for production of this material. For comparative purposes, EC-14 is a clinical isolate that caused disease in man. C-532 and the other two organisms are laboratory strains that, like many laboratory strains, have decreased pathogenicity. An $LD_{50}$ in normal animals shows at least a 600-fold difference in pathogenicity among these strains for the mouse. In the case of EC-14, if you want a different sort of measurement, this is about 0.5 ml of a 1:10,000 dilution of a 16-hour culture. The other strains would be dilutions something like 1:16 or 1:30.

In spite of this striking difference in pathogenicity, you will notice that the antibiotic sensitivity is essentially identical. I don't make a lot out of this; but I do use it to illustrate that one can assess the pathogenicity and antibiotic sensitivity of *E. coli* organisms used in a commercial-scale process. I am perfectly aware that they don't contain, to the best of my knowledge, any mammalian DNA, and I cannot assess whether that would have changed their sensitivity to antibiotics. But if and when such organisms exist, I believe we can make the appropriate determinations.

TABLE 3  Pathogenicity and Antibiotic Sensitivity of *E. coli* Strains

| | *E. coli* Strain | | | |
| | EC-14 | C-532 | C-532.5 | C-532.6 |
|---|---|---|---|---|
| $LD_{50}$[a] | $5 \times 10^4$ | $1.4 \times 10^7$ | $2.9 \times 10^7$ | $2.5 \times 10^7$ |
| Antibiotic sensitivity[b] | | | | |
|   Chloromycetin | $32^c$ | 64 | 64 | 64 |
|   Gentamicin | 0.5 | 0.25 | 0.5 | 0.25 |
|   Tobramycin | 0.5 | 0.5 | 0.5 | 0.5 |
|   Kanamycin | 2.0 | 2.0 | 1.0 | 2.0 |
|   Tetracycline | 8.0 | 8.0 | 8.0 | 8.0 |
|   Cephalothin | 2.0 | 4.0 | 4.0 | 4.0 |
|   Cephalexin | 4.0 | 8.0 | 4.0 | 4.0 |
|   Cephaloridine | 2.0 | 2.0 | 2.0 | 2.0 |
|   Cefamandole nafate | 2.0 | 2.0 | 2.0 | 2.0 |
|   Cefazolin | 0.25 | 0.5 | 0.5 | 0.25 |

[a]Approximate number of organisms per 0.5-ml dose (intraperitoneal) necessary to kill 50 percent of the mice in our standard mouse test.
[b]ICS agar dilution method for antibiotic susceptibility testing.
[c]Minimal inhibitory concentrations in micrograms per milliliter.

In summary, I have tried to indicate that insulin supplies are now adequate to meet the requirements of the diabetic population. I have also presented data suggesting the prudence and logic of at least entertaining the possibility that additional sources of insulin should be seriously considered, due to a number of unpredictable factors relative to gland supplies, and a predictable increase in insulin requirements. I think this contingency should be investigated slowly, carefully, and systematically. Eventually, however, there is the possibility that an additional controllable source of insulin may be required. This potential requirement is independent of the possibility of producing an insulin with improved therapeutic potential. I have suggested a number of other factors that may influence this potential requirement.

We can ill-afford to wait until an absolute shortage develops before initiating studies to develop these alternatives. I believe we have a good deal of time before that possibility happens, but we should use that time well. I believe the potential benefits to mankind of this type of research far outweigh the risks. The successful application of this technology to this and other problems would appear to me to be in the best interest of the scientific and medical communities as well as the patients and people of the world. This is truly "science for the people."

When this type of research is done, it should be for a reason. I am not in favor of willy-nilly shotgun experiments. I would not, for example, be in favor of putting echinoderm DNA into *E. coli*, as Dr. Sinsheimer suggests, unless there was a real benefit as an objective. But to patients who will die if they don't receive an essential hormone I see a benefit. I believe this is a subject that we ought to address ourselves to responsibly, without too much stridence, and in terms of its benefits to science, medicine, and society.

## REFERENCES

1. Diabetes Source Book, U.S. Department of Health, Education, and Welfare, Public Health Service. Revised 1968. PHS Publication No. 1168. Government Printing Office, Washington, D.C.

2. Babson, D. L. and Co. Staff Letter Reports 7/14/66.

3. Livestock and Meat Statistics, U.S. Department of Agriculture. 1970. Supplement 1975. Government Printing Office, Washington, D.C.

4. Dayhoff, M. O. (ed.). 1972. Atlas of Protein Sequence and Structure, vol. 5, pp. D206-D211. National Biomedical Research Foundation, Georgetown University Medical Center, Washington, D.C.

5. Guineé, P., N. Ugueto, and N. van Leeuwen. 1970. Appl. Microbiol. 20:531.

6. Hershfield, V., H. W. Boyer, C. Yanoksky, M. A. Lovett, and D. R. Helinski. 1974. Proc. Natl. Acad. Sci. U.S.A. 71:3455.

7. Crane, C. W., and G. R. W. N. Luntz. 1968. Diabetes 17:625.

8. Shichiri, M., A. Okada, K. Karasaki, R. Kawamuri, Y. Shigeta, and H. Abe. 1972. Diabetes 21:203.

## POTENTIAL RISKS

Ruth Hubbard

*Professor of Biology, Harvard University*

You will probably be surprised and perhaps pleased to find that Dr. Johnson and I are going to be very agreeable opponents. In fact, what I would like to do is essentially go on with the story that you have just heard, and raise some more questions about it, and to use it in a way as a case study of the kinds of issues that you have to look into when you try to make a decision as to potential benefits and risks of a new technology.

When it comes to assessing the risks of putting insulin genes into *E. coli,* I really don't want to spend my time trying to dream up scenarios of what might or might not go wrong in the technical sense. We know too little about the technological hazards, as has been pointed out before, and added to that we know too little, in fact almost nothing, about how insulin works. So to start dreaming about what will happen to *E. coli* that have active insulin inside them is completely futile, because we don't know whether insulin gets into cells, what it does if it gets into cells, what insulin fragments do in cells, so I think this would be a waste of time.

It is clear that if we indeed developed little insulin factories in our guts that produced insulin out of control that this might be a bad thing, but we all agree that that is relatively unlikely, although of course it is possible. So what I would rather concentrate on are the hazards of looking for technological fixes as solutions to complicated diseases of metabolic control and other complicated problems. Those of you who follow this field will probably recall that somewhere between a couple of years and six months ago when the challenges to this technology first went up, we were quoted benefits that ranged all the way from curing world hunger to curing cancer, curing diabetes, curing a whole variety of other ills.

Ethan Signer tomorrow, I believe, will try to allay the misconception that this technology will feed the starving millions. Curing cancer hasn't been talked about much lately, because I think a number of people have become cognizant of the fact that the technology is at least as likely, if not more likely, to produce cancer than to cure it.

Only six months ago we were still hearing about how this technology would allow us to put insulin genes into diabetics. We haven't heard much about that lately, probably because the people who were saying this were clued in by their colleagues to the fact that there is no reason to believe that any diabetic lacks the gene for producing insulin. Diabetes is a disease of control; it is not a disease of lack of insulin.

Now, the fact is that juvenile diabetics do in fact stop producing insulin, but that does not mean that they don't have the gene for producing it.

So let me now go into the disease of diabetes and dissect it a little more than was done in the previous talk. There are different estimates of the number of diabetics in this country. The estimates depend in part on how you define a diabetic. If you define as diabetic a person with a certain level of blood glucose, then there are a great many more diabetics than if you define a diabetic as a person who feels ill and comes to the doctor or the hospital with symptoms. But the estimates vary somewhere between 2.5 and 5 million in this country. Of these about 80,000 are so-called juvenile diabetics: persons who become diabetic before the age of 17, have a very rapid and catastrophic onset of the disease, and very quickly lose the ability to produce insulin in their pancreas, and who absolutely need insulin in order to survive.

Now what causes juvenile diabetes?  It is surprising how little we know about this more than fifty years after the discovery of insulin, and most people not familiar with the field, I think, overestimate what is in fact known about it.  The genetics of diabetes is a very hazy area.  A decade or so ago medical textbooks were telling us that adult-onset diabetics had one gene for diabetes and juvenile diabetics had the double dose.  Nobody believes this any more.  If there is a genetics of diabetes it probably involves a great many genes.  There certainly are very large environmental contributing factors, and what, if any, role genetics plays is not clear.

One of the circumstances that is often associated with the onset of juvenile diabetes and is implicated by some as a cause is a number of viral infections.  Concerning the genetic link we find that identical twins among juvenile diabetics have somewhat less than 50 percent coincidence of the disease.  And, interestingly enough, it turns out that if the second twin does not get diabetes within about a year of the first one-- these are identical twins--chances are she or he never will.  So this again suggests that there may be an environmental component, perhaps related to the viral infection.

All right, so clearly juvenile diabetics, about 80,000 of them, need insulin.  The cost of insulin, by the way, was not cited by the previous speaker, but it has in the past been cited as one reason why we should develop the recombinant DNA technology to produce it.  Interestingly enough, at present the cost of insulin to a juvenile diabetic who abso- lutely depends upon it is about ten cents a day, and about twelve cents a day is spent on throw-away syringes.  So if we are going to try to cut costs, one of the possibilities is to teach people again how to sterilize syringes so that we don't have to use throw-away syringes any more.

Now the vast majority of diabetics, 95 percent or more, are what is called adult-onset diabetics, people who develop diabetes above the age of forty or forty-five, develop it slowly, and many of whom, we are told, do not even know that they have the disease, whatever that means.  (Is "disease" a subjective state of being or an extremely defined one?)

There are a number of interesting epidemiological facts about adult- onset diabetes that I think are important to bear in mind when you think about trying to cope with this problem as a medical problem and not as a scientific problem.  So, for example, the mortality of whites from diabetes in this country has remained more or less steady over the last twenty to twenty-five years, but the mortality of nonwhites has increased steeply from about the same level as that for whites in the early 1950s to a present level about twice that of whites, 25 versus 14 per 100,000. And this increase has been steady and continuous.  These data are from an HEW report published in 1974, and neither in it nor anywhere else have I seen any suggested reasons for this increase.  The mortality of nonwhite women from diabetes is almost 2.5 times that of white women, and again I have not been able to find any suggested reasons.

The prevalence of diabetes in people living in families whose heads have less than nine years of education is about three times that of those in families with heads that have more than thirteen years of education. The prevalence of diabetes among people with family incomes less than $3,000 a year is almost four times that of people with family incomes greater than $15,000 a year.

What I am suggesting is that what we need to know in order to study the cure for diabetes are the causes of diabetes, which are, as with all other diseases, heavily influenced by social and environmental factors. This is not to downgrade diabetes as a health problem. It obviously is; it is among the top eight killers in this country. But we need to know more about its real causes, and the real causes are not lack of insulin. The fact is, as many of you probably know, that most diabetics secrete more insulin than do normal people. Many develop an insulin "insensitivity." There are more than nine possible ways in which the carbohydrate metabolism of diabetics may malfunction; I am not going to bore you with the details. But the derangement of the metabolic controls is obviously potentiated by a large number of factors, the most tangible and generally agreed upon of which are a high-calorie diet, lack of exercise, and obesity.

Insulin cannot cure diabetes. It can control the extreme fluctuations in blood sugar, but it is not clear that controlling these extreme fluctuations has any effect on the long-term progress of the disease, which usually ends in death, in most cases, through vascular diseases of one sort or another.

A few years ago, in 1970, a large study conducted by twelve universities known as the University Group Diabetes Program was completed. It was an eight and a half year study that tried to randomize persons with adult-onset diabetes and expose them to four types of therapy. One was oral antidiabetic drugs; one was a constant, unmodified dose of insulin, based on their total body surface; another was a variable dose of insulin based on their blood-sugar levels; and the fourth group was receiving a placebo. These were all people whose diabetes could be controlled through diet, which depending on whom you listen to, is either a small fraction or the vast majority of adult-onset diabetics. (There are in fact doctors who believe that the proper therapy for adult-onset diabetes, in most instances, is a low-calorie diet and exercise, so keeping the weight down to normal limits.)

Figure 19 depicts a study done with 823 patients. It was discontinued after eight and a half years because it showed that the persons who were getting the oral antidiabetic drug, Tolbutamide or Orinase, had almost double the cumulative death rate of the others. This was so for deaths from various vascular diseases; but deaths from all causes also were considerably higher in the Tolbutamide group, although the differential was not quite as great. But the thing I want to point out to you here is that the group on the steady dose of insulin, the one that is labeled ISTD, the group on the variable dose of insulin, and the group on placebo were essentially indistinguishable.

So at present we have in fact no way of assessing the demand for insulin, because the insulin therapy that many diabetics receive is challenged by many doctors as being not only unnecessary, but wrong,

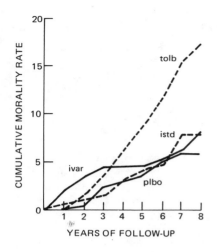

FIGURE 19 Cumulative mortality rates per
100 population by year of follow-up.
TOLB, Tolbutamide; IVAR, variable insulin;
ISTD, standard insulin; PLBO, placebo.
(University Group Diabetes Program.)

because it discourages people from going after the basic problem, which
is obesity. And there the only "cure" is to cut back on calories,
increase exercise, and so keep the weight down.

The study that I just referred to raised a considerable hue and cry,
particularly among the manufacturers of oral antidiabetic drugs, and
it was therefore reevaluated in 1975, and, interestingly enough, the
report of the committee that reevaluated it, published in the *Journal
of the American Medical Association* in 1975, concluded among other
things that what is needed now, more than fifty years after the discovery
of insulin, is a similar study to evaluate the efficacy of insulin.

So the thought I want to leave you with is this: before we jump
at technological gimmicks to cure complicated diseases, we first have
to know what causes the diseases, we have to know how the therapy that
we are being told is needed works, we have to know what fraction of people really need it. There are lots of questions that we have to answer
in order to lick diabetes, and diabetes is a major health problem. But
what we don't need right now is a new, potentially hazardous technology
for producing insulin that will profit only the people who are producing
it. And given the history of drug therapy in relation to other diseases,
we know that if we produce more insulin, more insulin will be used,
whether diabetics need it or not.

## DISCUSSION

JOHN ITIALIS, National Health Federation:  I would like to make a couple
of comments on the meeting held here for the last two days, to the
people in the front of the bus as compared to the people in the back

of the bus.  As may have been implied, I don't think there is that
much intelligence separating the two groups.  As a matter of fact, I
think there is plenty of ignorance to go around, and this has been
displayed pretty liberally at this Forum.

I am not saying this disrespectfully.  I think that scientists
and lay people both have a lot to learn about the so-called secrets
of the genome, about the benefits as well as the risks of this
recombinant DNA research.  I would like to note a couple of the
points that have been made that I am very sorry to hear from both
sides.

First of all, I think I know enough as a scientist to reassure
those lay people who feel that we are going to manipulate human genes
to the extent that we are going to have super races or good laborers
or subtle workers.  We don't have this technology.  I don't see it
in the foreseeable future like maybe some of the people like Lederberg,
etc., were quoted as seeing.

I also disagree with the idea of people who think we are going to
cure cancer and heart disease by means of recombinant experiments
and manipulation of human genes.  You take the heart, for example.
Just look, take common sense and look.  If I am a patient with heart
disease, even if I have something, even if I have a method for getting
DNA into a cell to uncode or code or do something that is going to
improve my heart genetics, what am I going to do with that cell?  Am
I going to remove every cell of my heart out and then recombine it
with DNA and then put the cells back together and then put a heart
in?  Where am I going to be in the meantime?

How about a thalassemic?  Are we going to drain his marrow out,
then culture his cells, get DNA in and put it back in?  Quite frankly,
I would rather be a thalassemic than have that happen to me.

So these are the problems, I think, that we are involved in.  We
have been promised great benefits, and they don't exist.  We have
these great hazards which have been brought up and for which we don't
have the technology and the knowledge, thank God, to implement.  So
I think what we are left with, primarily, is the accidental release
of these recombinant DNA products into the environment.  Many people
have mentioned how *E. coli* K-12 is a harmless type thing.  Well, what
happens if we recombine enough of its genetic material to make it
less harmless?  There are so many hazards involved, I think, with
this thing getting out into the environment that we don't know about
that we just can't predict this so-called risk versus benefit.

The second aspect I would like to address myself to is once we
finally do get a recombinant that has been shown to be agriculturally
beneficial, the decision must then come to purposely take this out
of the P4 lab and put it into the environment.  I would hate to leave
it up to the EPA to decide whether or not this is going to be good for
our environment.

I would like to just raise one little question.  Maybe it is a
scare tactic, but what would happen if we had a group of scientists
walk into a P4 lab one day, and 5:00 o'clock came around and no one

came out.  And at 6:00 o'clock no one came out.  And at 7:00, 8:00,
9:00, and 10:00 o'clock no one came out.  Who would go in there after
them?

FRED WHITEHOUSE, Internist, Henry Ford Hospital, Detroit:  My practice
is specializing in diabetes, and I follow about 1,500 patients, about
two-thirds to three-quarters of which are diabetic.  I also am an
officer of the American Diabetes Association, which is a voluntary
health agency aimed at helping the diabetics' lot.  I would like to
congratulate both of the speakers in the succinct remarks they made,
and also to make one or two observations.

     The American Diabetes Association and most of us who are respon-
sible for advising diabetics recognize that these people are human
beings first and diabetics second.  And therefore we hope officially
as well as informally and personally that much of the good that will
come out of this Forum will be for the good of human beings.  Then
if something else is left over for the diabetic, we will be delighted.

     There are some remarks here relative to the comments from Drs.
Johnson and Hubbard that need a little bit of clarification, as I see
it, as a practicing physician.  That is that there may be a race
on between the cure for diabetes, if that is found, and the running
out of insulin supplies, as Dr. Johnson might have implied.  And it
is true, as Dr. Hubbard mentions, that many of the diabetics are
adult-onset and obese.  But pragmatically at the present time, try
as we will, we find it very difficult to get these people to the point
where we can control their diabetes well with diet alone.

     I would like also to distinguish between juvenile-onset diabetes
and insulin-dependent diabetes.  Many of the juvenile-onset diabetics
are insulin-dependent, 99 percent.  But many of the adult-onset people
become insulin-dependent because of their elevated blood sugars for
whatever reason, and sometimes they develop life-threatening situations
related to acute complications.  So it is a bit of an empty argument to
say that there are just two clearcut types of diabetics.  Many will
go from one to the other, and there is this problem of the continued
need for insulin.

     In addition, there are situations that do prevail that make it some-
what evident at the present time that there may be a pertinence to
the control of diabetics vis à vis blood-vessel complications, particularly in microvascular disease.  So I am pleased that a problem
disease, diabetes, has been discussed at a Forum like this, and we
hope that there will be positive things coming out of this that will
permit further advance in knowledge that will help not only the
diabetics, but human beings generally.

MEREDETH TURSHEN, Oil, Chemical and Atomic Workers International Union:
I would like to make a very brief comment on the proceedings this
afternoon.  I would like to compliment and thank Dr. Hubbard for her
effort to present a very clear case, to present it in words that were
not mystifying to the public here, in ways that we could all follow

her discussion and understand what her point was. This has not been true of many other speakers. I think that we would like to express our appreciation to the Academy for making this a public forum. We would like to say that Mr. Mazzocchi this afternoon presented very cogently, in fact he documented the reasons, why public participation in this Forum is so important.

It seems to me that speakers like Professor Davis, in their disdaining remarks, which I found very insulting, were not disguised by the bland style of their delivery, and, in fact, are saying that public participation is a nuisance. In calling for further forums and very clearly enumerating the sorts of people he wanted to dialogue with, he left out the public.

DeWITT STETTEN, NIH: I would like to make a comment and raise a question. The comment relates to the fact that I am perhaps older than some of the people in this audience, old enough to remember that there was an impending or threatened insulin shortage in Indianapolis in the years immediately following World War II. I happen to know this because I was one of the fortunate persons to be invited, year after year, to an interesting colloquium held at Eli Lilly in the years following the war, when the reserves of insulin, of which I believe at that time Eli Lilly tried to preserve an eighteen-month supply, started to decline. And when they got down to about a six-month reserve, Eli Lilly became extremely anxious. The reasons for the decline were related, I believe, to the rationing of beef and to the decline in the cattle industry as a result of the war effort, and to the increase in black-market slaughtering; black-market pancreases were not available to the industry.

By all odds I think we are perhaps in a more perilous position with respect to insulin supply than Dr. Johnson's remarks led me to believe, and I would be interested in his comment. I also have a question. It would be my guess, though I have no certain knowledge, that the protein for which the beta cell of the island codes is not insulin, but is its precursor, proinsulin, preproinsulin. Preproinsulin, as far as I am aware, is an agent of no pharmacologic activity.

I would like to inquire whether this is true, and whether this doesn't change the argument somewhat; and how easy and facile is the laboratory conversion of preproinsulin into a pharmacologically useful product?

JOHNSON: I think all your comments are quite accurate. The thing that you are really interested in is proinsulin, and you are quite accurate that it is pharmacologically inactive, or essentially so. It is at least much less potent than insulin. It is relatively easily converted to insulin by proteolytic enzymes. There is now, of course, as there is in all the hormones, the possibility of a preprohormone; I don't know whether there is a prepreprohormone or not. But it seems to be true that all of the protein and peptide hormones are derived from a larger structure.

I would like to make one other statement since you brought up the colloquium that you used to attend. We in industry are occasionally accused of working behind shrouds, or something of this sort. As we do every now and then, in May of 1976 we had a symposium on insulin biosynthesis and the possible applications of recombinant DNA technology. This was a meeting which was by invitation, but people were invited from almost all over the world. I personally wrote a letter to the editor of *Science* and suggested that he might like to send Nicholas Wade or some other *Science* writer to review and report on this meeting. This invitation was not accepted; however, we did summarize the meeting, and I reported it in the Nucleic Acid Recombinant Science Memo of NIH, along with a statement about how we felt about the guidelines for this type of research.

If I remember rightly, there was even a forbidden experiment reported there, not from us but from one of the academic participants. But it is all there so that people can look at it and see what we were talking about and what we were interested in.

HUBBARD: I want to make one comment on the pharmacological activity. That is, if proinsulin were to get into *E. coli*, presumably *E. coli* would, or is very likely to, have the enzymes to convert proinsulin into insulin. In fact we don't even know, and that is what I was trying to refer to, though very superficially, that insulin itself is not a proinsulin. We don't know what, if any, component of insulin gets from outside of the cell membrane into cells, and we therefore don't know whether *E. coli* could convert proinsulin or insulin into a more pharmacologically active species than insulin itself.

JOHNSON: Well, if you further break down insulin you lose the whole biological activity.

WILLIAM REZNIKOFF, University of Wisconsin: I want to share a few observations I have about the meeting, and that is basically I think we have been discussing the issue of recombinant DNA on two levels, although it hasn't been specifically stated as such. I think this is important, because on one level I am not sure we can ever come to consensus; on the other level I think we have come to consensus; and I think it would be a shame that we leave the meeting tomorrow and not act on that consensus.

The first level is a question of recombinant DNA as a basic research tool. We don't have a consensus of that. Some people feel that it is essentially a uniquely moral type of problem, which I don't happen to agree with, but I acknowledge the difference of opinion, and also about the immediate dangers of biohazards to the basic research work.

What I do feel that we have a consensus on is the question of the applied use of recombinant DNA work, or any other basic science work. I feel that we have come to a feeling that it should be in some manner regulated in an open forum, with inputs not just from scientists, but also from workers and epidemiologists and people who are outside the immediate scientific community.

I think that we should have a public organization that should be involved with an initial analysis of whether a particular applied use of basic research work, and in this case, recombinant DNA work, has significant dangers as opposed to its proposed benefits. There should be follow-up analysis of the applied work.

Almost every speaker here has pointed to the difficulties of applied problems, the difficulties of whether insulin is an economical and a useful applied utilization of recombinant DNA technology. The question of genetic engineering of human beings, which I don't think is relevant to recombinant DNA work, since we have other classical techniques of human genetic engineering, again is an applied problem. We have to be clear that these applied problems need to be addressed, and I think it would be irresponsible of us to be frozen in a debate about the value of the basic research work, because I don't think we are going to come to a consensus about that. But we can come to a consensus about the importance of evaluating the applied work.

STANLEY COHEN: As some of you may know, in addition to being a worker in the field of recombinant DNA, I am a physician, and I teach clinical medicine and clinical pharmacology at Stanford. And as a physician and as an internist I can't let some of the statements that Dr. Hubbard has made, which are just simply not correct, go unchallenged. The implication of some of her comments seems to be that there may be a lot of diabetics around, but that these can be treated by diet alone if we could just get them to reduce. That is only true for some. Certainly, there are a certain number of individuals commonly known as obese diabetics that can be managed by diet alone, but there are many other diabetics that cannot be managed by diet alone.

Now, there is another point that Dr. Hubbard made that I would like to question, and that was her statement that *E. coli* probably would have the enzymes to cleave the first few amino acids from preproinsulin or from proinsulin, resulting in an active hormone fragment which would have the dangerous consequences that she is worried about.

It seems a little strange to me to have this statement made, after we have been listening to Dr. Hubbard and Dr. Wald say that it would be *unnatural* for *E. coli* to have enzymes that perform functions similar to those in eukaryotic cells. Yet at this moment Dr. Hubbard seems to be asserting that the very specific enzymes that are required for the posttranslational processing of proinsulin to insulin would be likely to be present in *E. coli* just by chance. If *E. coli* does have the same range of enzymes as eukaryotic cells, it is hard to see how the introduction of these eukaryote genes would impart truly novel properties to the bacteria. There appears to be a contradiction here. I wonder if you would be prepared to comment on either of those points.

HUBBARD: In the course of this brief statement you have misrephrased, I believe, four statements of mine, which were the only four you quoted. Fortunately we do have a record of what was said. In fact, you haven't even misphrased them. It is the usual business of the music making the sense. I mean they were very simple statements like most or many diabetics can be controlled by diet, I say. You say that I said almost all diabetics can be controlled by diet.

COHEN: The implication and thrust of your presentation, Dr. Hubbard, seemed to be that insulin is really not necessary, at least--

HUBBARD: No, the implication was that we do not know at this point what the need is.

COHEN: Oh, I see.

HUBBARD: We do not know what the epidemiology of the disease is. We do not know the extent to which insulin is needed.

COHEN: I would guess that most physicians who have treated persons in diabetic coma, diabetics with infectious diseases that have severe problems, and others that require insulin would disagree with your statement. For many, insulin is a life-saving agent--

HUBBARD: For those who need it.

COHEN: That is right. For those who need it.

HUBBARD: Correct.

COHEN: Thank you. Then we agree on that. Perhaps you could comment on the second point, your statement that it is likely that *E. coli* would have the enzymes to chop off--

JONATHAN KING: No, that there *might* be enzymes.

HUBBARD: I believe that *E. coli* has intracellular proteases. Are we agreed on that?

COHEN: We are talking here about some very specific posttranslational processing of a complex molecule, not just some random protease activity.

KING: I would like to make a comment about the structure, the kind of intellectual structure of this argument about how we are going to make insulin, just to show that it is not an intellectual argument at all.

We are told that we could make insulin in *E. coli*. It is recognized that there are many barriers. There are extraordinary barriers: we need P4 facilities; we need ingenious molecular biologists. But we can overcome those barriers, and therefore we must.

On the other hand we are told that we are running out of cows or pigs. When we suggest let us grow some more pigs, no, no, there is going to be a drought, and we are not going to be able to grow any more pigs. It is extraordinary that people can stand up there, lay out an extraordinary series of technological barriers, and say we can and must overcome them to produce insulin that way. But no, we can't put a little more public investment into the cattle industry.

Now, I submit to you, if standing up earlier today on the podium would be representatives of the American Cattlemen's Association and the American Breeders' Association, that they would give you a very, very different assessment of the problems of getting enough insulin out of wildlife. Besides that, you know, you can develop a technology to culture the cells and make it. As a matter of fact insulin, of course, has been synthesized *in vitro*, not only by us but by the Chinese, and it is done routinely experimentally to study amino acid substitutions.

We are told that it is very hard to make purified insulin *in vitro* synthetically. It is not hard to make it in *E. coli*; there are only thirty-seven steps we have to solve there.

Essentially here you have an argument that really comes from the fact, well, my business is doing recombinant DNA and so I offer that as the solution to the insulin problem. If we had people here whose business was growing cows they would offer that as the solution to the insulin problem, and the guys who work at Rockefeller Institute who do it synthetically would argue that theirs is the way to make insulin. These are not scientific imperatives, these are social choices, absolutely and completely, and there should be no confusion about it whatsoever.

JOHNSON: I think I pointed out that I don't cause drought, and Eli Lilly doesn't cause drought. I said that we don't have control over environmental factors, and I think that is true. I pointed out that there were a number of other alternatives which are being pursued, both at Lilly and at other laboratories. I did not say that we could make insulin with *E. coli*. I don't know that. In fact, the people at Harvard probably have a better concept of that than I do, because they are much closer in terms of inserting the gene for insulin than we are.

I think some of these statements that Dr. King has made are some-what irrational. For example, swine and cattle are not raised for their insulin potential. Pancreas is strictly a by-product of the meat industry. A pancreas from a steer with a value of over $400 at today's prices has a value of approximately 75 cents. Doubling, tripling, or even increasing the price of pancreas by a factor of ten wouldn't have a substantial impact on the number of cattle and swine raised, but it would have a drastic impact on the cost of insulin to the diabetic. The demand for and the price of red meat and the prices and availability of feed grains are the primary factors determining the numbers of cattle and swine raised and I doubt that any spokesman for the cattle or swine industries would tell us otherwise.

Dr. King also states that there are a number of technological barriers, I believe he stated thirty-seven, to overcome before production of insulin by *E. coli* or some other microorganism can be achieved. While I am not sure of the exact number of problems to be solved, I would certainly agree there are many problems to be solved, but I would also point out there are significant problems to the other potential sources of insulin I mentioned. The beta cell does not grow well in culture, and while the chemical synthesis of insulin has been achieved after years of intensive laboratory effort, such a synthesis involves over 200 separate and distinct chemical reactions. Even if one assumes high yields at each of the 200 steps of synthesis, the cumulative yield losses are large and the yield of end product is small. We believe the costs of such a product would be prohibitive. We are in the business of trying to produce useful, life-saving medications, and what we are concerned about is whether we can do that.

KING: I apologize. I didn't mean to imply that Eli Lilly Company either has or should have complete control over insulin production in the United States. As far as I am concerned it is the people of the United States who should have control over that process. And even though Eli Lilly may not be able to control that, for example, you can easily get into a situation where the return on investment from producing it, getting it from cows, just wouldn't be worth your while to do it, right? But that doesn't mean that we don't have--

JOHNSON: That doesn't mean that we wouldn't continue to do it. We have produced medicines at cost, Dr. King, because there is some medical need for them. We have produced a great many we have lost money on. If there is a medical need, somehow we will try to meet it.

My real expertise is in cancer chemotherapy, although I am a cell biologist by training, and we isolated a drug which is useful in the treatment of several forms of cancer, including leukemia, in which you get an ounce out of a ton of dried leaves from a plant grown in Madagascar. Initially we didn't make money on that. We provided the drug at cost because there was clinical evidence that it was useful in the treatment of disease.

SHELDON KRIMSKY, Tufts University: I would like to ask Dr. Johnson whether Eli Lilly is engaging in recombinant techniques now, what are its plans, and what is its posture toward the possibility of further regulation and the present guidelines, and what plans does it have for lobbying against further regulations.

JOHNSON:  Well, let me answer the last one first.  We don't have any plans to lobby against it.  We are engaged in some forms of recombinant DNA research.  These are fairly general, and primarily are involved in its potential application to the fermentation area.  It is obvious, I think, that if one could clone out a gene that controlled the metabolic pathway for a certain metabolite which had medical usefulness that one might be able to produce it more economically because you could increase yields.  It is possible, of course, that one might be able to produce different kinds of metabolites with medical application.

Now, I have made no secret of the fact that we are interested in the possibility of the theoretical application of this to compounds like insulin.  I mentioned, and maybe you didn't hear me, that we discussed this in an NARSM report.  In that same report I said that the policy--I couldn't speak for the industry, I could only speak for Lilly--but I said that we thoroughly endorsed the NIH guidelines. I have made every effort, both within Lilly and within the Pharmaceutical Manufacturers Association, to lead the industry toward complete and voluntary compliance.

Every experiment we do is in complete compliance with the NIH guidelines.  It appears inevitable that there will be legislation in this area.  I would like to suggest to you that that is something that we are going to be able to live with a lot more comfortably than you.  In industry we are used to having people stand over our shoulder and suggest what we do, and I don't think that you in academia are, and I don't think you are going to like it.

# DAY III

ALEXANDER RICH
*Cochairman*

DAVID HAMBURG
*Cochairman*

# PRIORITIES FOR DAY III

Alexander Rich

Yesterday we had three of our case analyses dealing with different aspects of research on recombinant DNA. Today we will continue this and explore in some detail another case analysis. Last evening most of you participated in a variety of workshops. These were designed to give us an opportunity to explore in considerable depth subjects that we could only cover peripherally or not at all in the plenary sessions. We will start today with reports from the chairmen of these various workshops. To present these I would like to call on Dr. Hamburg.

DAVID HAMBURG: I want very much indeed to thank the people who participated in the workshops last night. I really am impressed with the dedication and thoughtfulness and constructive work that went into them. I particularly want to thank the chairmen and cochairmen, who nc only ran the meetings effectively, saw to it that all points of view were well represented, but have written the reports that follow.

REPORT OF WORKSHOP NO. 1
Is It Likely That *E. coli* Can Become a Pathogen?

*Cochairpersons:*   Richard Goldstein
*Assistant Professor, Department of Microbiology and
Molecular Genetics, Harvard Medical School*

Elena O. Nightingale
*Senior Professional Associate, Institute of Medicine-
National Academy of Sciences*

The title of this workshop should have been "What are the issues to consider in the use of *E. coli* as a cloning vehicle?" One of these issues
is "Is it likely that *E. coli* K-12 can become a pathogen?" Numerous
types of *Escherichia coli* are known to cause primary disease in the
gastrointestinal tract of man and of many animals. *E. coli* are also
known to behave opportunistically and to cause disease in other body
sites such as the urinary tract. Further, pathogenicity* of *E. coli*
and of enteric species in general is a very complex process, some parts
of which are not yet understood. Usually, pathogenicity cannot be attributed to a single determinant, but is a reflection of a constellation
of bacterial genes, some chromosomal and some extrachromosomal, acting
in concert with a multitude of specific and relatively nonspecific animal
host factors. However, more important than the pathogenic potential

*In this discussion, pathogenicity is defined as the ability to cause
disease or to interfere with normal physiological function.

of *E. coli* is the concern about other bacteria that may receive foreign genetic information from the organism in question.

This workshop was attended by twenty-five to thirty people, including several experts in enteric diseases; some general clinicians; basic scientists, including molecular biologists, microbiologists, and geneticists; and some members of the public. The workshop goals were to have some of the facts about *E. coli* and about laboratory-acquired infections presented by experts in these fields, followed by open discussion of a full spectrum of the thoughts and major themes of this area. In spite of the expected divergences of opinion, the workshop discussions were balanced and thoughtful and proceeded to the proposal of some experiments to provide a data base for risk assessment in the use of *E. coli* strains in recombinant DNA research.

Dr. H. Bruce Dull, Assistant Director for Program, Center for Disease Control, began the program by presenting some of the data that have been collected on laboratory-acquired and laboratory-associated infections. Dr. Dull emphasized that even using the best containment, there has been an inevitable but fortunately small number of laboratory-associated or laboratory-acquired infections. The risk appears to be related most directly to the behavior of people in the laboratory or to the lack of consistent behavior rather than to a documentable accident, such as breakage or spillage. The recent deaths of two janitorial workers at the Center for Disease Control in Atlanta provide a case in point. The deaths were caused by the rickettsiae of Rocky Mountain spotted fever. These workers were exposed to an aerosol of the rickettsiae in a still unexplained manner. Dr. Dull emphasized that there are no risk-free laboratories, that some laboratory-acquired infections will occur whether by accident or because of human behavior regardless of precautions and containment. It is therefore important to plan in advance for such occurrences by developing whatever controls are needed, and by education and tying in of related health services.

Dr. Samuel B. Formal, Chief, Department of Applied Immunology, Walter Reed Army Institute of Research, summarized some of the characteristics of enteric pathogens. Enteric pathogens can cause disease by two major mechanisms. One is by the ability to multiply in the small intestine and elaborate an enterotoxin (an example of this type of pathogen is *Vibrio cholera*), and the second is by the ability of an organism to invade the epithelial cells of the colon, and possibly of the ileum, multiply in the epithelial cells, and cause ulcerative lesions. The shigellae which cause dysentery and some forms of salmonellae cause disease by this mechanism. Certain strains of *E. coli* can cause a choleralike disease and other strains can cause a dysenterylike disease. Experiments done in swine showed that both of these factors are controlled by plasmids.[1] It is possible to construct strains of *E. coli* with either of these plasmids, or with both of them. The organism with just the adhesive factor produces mild diarrhea in piglets; if only the toxigenic factor is present, no diarrhea resulted. If both the toxigenic and the adhesive factors were present, severe diarrhea developed. Of interest are the observations that if *E. coli* K-12 was substituted as the recipient strain for these plasmids, no evidence of disease resulted. Further, although it has been reported that *E. coli* K-12 is unable to colonize[†] the bowel when fed to healthy well-nourished mammals of various kinds or humans, some *E. coli* K-12 cells can survive passage through the intestinal tract.[2,3] The inability of *E. coli* K-12 to colonize the normal intestinal tract is probably related to the fact that *E. coli* K-12 is defective in the production of lipopolysaccharide, which is a component of the capsule. However, *E. coli* K-12 could colonize the intestinal tracts of individuals whose normal flora had been disturbed due to disease, fasting, or antibiotic therapy.

The limited experiments done so far with *E. coli* K-12 fed to human volunteers do not demonstrate pathogenicity or colonization of the human gut by this type of *E. coli*. But, because of the large variation of results among individuals or within an individual with time for interaction of enteric organisms with the human gut, consensus was reached on the desirability of feeding significantly large numbers of volunteer subjects, both male and female, with *E. coli* K-12 in varying doses and studying the fate of these organisms, not just by culturing feces, but also by culturing urine, perineum, and pharynx. Discussion of the pathogenicity of *E. coli* and the potential pathogenicity of *E. coli* K-12 for the bowel and for the urinary tract emphasized that relevant data are lacking on these topics. It is not clear what roles the adhesive factor or toxin play in initiating human urinary tract infections. Since at least some recurrent urinary tract infections in man are caused by fecal bacteria,[4] passage of *E. coli* in feces may be significant. Therefore, it is important not to ignore extraintestinal sites in studying the survival and invasiveness of any *E. coli* strains in man. Additional data on the interaction of *E. coli* K-12 with the human gut and with selected other body sites and particularly data on such interactions of K-12 candidate strains for recombination experiments would also be helpful in assessing risks in the use of these bacteria as hosts.

The workshop participants also agreed that the use of antibiotic resistance markers in recombinant studies should be avoided, since such markers would facilitate survival of the DNA fragments to which they are attached and of the host which bears them. Other markers useful for selection such as colicin and phage resistance are already available, and their use should be encouraged. It was also recommended that experiments be done soon with volunteer subjects on the mobilization of "nontransmissible" plasmids by various different factors. It is the fate of the recombinant DNA attached to a plasmid, i.e., any of the more than forty species of bacteria to which it may be transferred, which is of far greater concern than the pathogenic potential of the *E. coli* host bacterium itself.

Agreement was reached on the necessity of minimizing contact of persons with conditions such as decreased stomach acidity, poor immune defense, or any compromising conditions with laboratories conducting DNA

---

*Unpublished observations reported by S. B. Formal.
[†]Persist for seven or more days.

recombinant research with types of *E. coli*. The location of containment facilities for recombinant DNA research especially in relation to hospital patients should be given serious consideration.

Some of the participants wished to emphasize that *E. coli* K-12 is being used in DNA recombinant research because after three decades of experience with this organism no known disease has emerged, because the years of experience have provided intimate knowledge of the genetics and biology of this type of bacterium, and because such knowledge enables understanding and control of the experimental situation. Others felt that these reasons for continued use of *E. coli* K-12 do not suffice because of the ubiquity of *Escherichia coli* and the readiness with which it exchanges genetic information with so many other species of bacteria. Everyone felt, however, that the search for another bacterium that does not have intimate association with man and animals should continue.

Because the concerns about the pathogenicity of *E. coli* are extrapolated to include *E. coli* K-12, which differs in many essential respects from *E. coli* strains recently isolated from animal and human disease, the participants agreed that a clearly written article on current knowledge about and comparison of the ecologic, genetic, physiologic, pathogenic, and other characteristics of wild-type *E. coli* strains and of *E. coli* K-12 and its derivatives should be available to the general reader. Perhaps such an article could be written for *Scientific American, Science,* or *Nature.* Such an article might facilitate the dialogue among persons with different backgrounds and points of view on this issue.

The workshop participants were hopeful that the NIH-sponsored conference on the biology of *Escherichia coli* and on risk assessment of working with *E. coli,* to be held in Massachusetts in June of 1977, would help to delineate some of the issues discussed at this workshop further and might also result in a useful publication on the ecology and biology of *E. coli* types.* Finally, it was emphasized by the enteric disease experts present that the major issue was not the pathogenic potential of *E. coli* K-12 or its derivatives but the effect that new genetic material might confer on the properties of these organisms and on the microorganisms with which they interact. In every experiment, that new piece of information is the "catch-22."

REFERENCES

1.  Smith, H. Williams, and M. A. Linggood. 1971. Observations on the pathogenic properties of the K-88, HLY and ENT plasmids of *Escherichia coli* with particular reference to porcine diarrhoea. J. Med. Microbiol. 4:467-485.

*The ecology of *Escherichia coli* was reviewed for the members of the NIH Recombinant DNA Molecule Program Advisory Committee by Stanley Falkow in 1976. This document has not been published.

2.  Smith, H. Williams. 1975. Survival of orally administered *E. coli* K12 in alimentary tract of man. Nature 255:500-502.
3.  Anderson, E. S. 1975. Viability of, and transfer of a plasmid from, *E. coli* K12 in the human intestine. Nature 255:502-504.
4.  Levy, S. B. 1977. Fecal flora in recurrent urinary-tract infection. New Engl. J. Med. 296:813-814.


ADDITIONAL REFERENCES

Formal, S. B., P. Gemski, Jr., L. S. Baron, and E. H. LaBrec. 1970. Genetic transfer of *Shigella flexneri* antigens to *Escherichia coli* K12. Infect. Immun. 3:279-287.

Falkow, S. 1975. *Infectious Multiple Drug Resistance*. Pion Limited, London. 300 pp.

So, M., H. W. Boyer, M. Betlach, and S. Falkow. 1976. Molecular cloning of an *Escherichia coli* plasmid determinant that encodes for the production of heat-stable enterotoxin. J. Bacteriol. 128:463-472.

REPORT OF WORKSHOP NO. 2
How Can We Assess the Benefits and Risks of This Research?

*Chairman:*   William W. Lowrance
              *Special Assistant to the Under Secretary of State for*
              *Security Assistance, Science and Technology*


For working purposes the group took "risk" to be a quantity compounded of the probability of harmful effect and the magnitude of the consequences of that effect. Benefit was defined parallel to that.

It was agreed that the public decisions over such an issue as recombinant DNA research have three aspects: empirical assessment of the harmful and beneficial effects, normative appraisal of the social value (good or bad) of those effects, and sociopolitical policymaking. These three aspects are not totally independent, but it is helpful to distinguish them where possible.

It was pointed out that the kinds of risks most people are concerned about in this research are those of possibly extraordinarily important consequences but having very low probability. There are strong parallels to the early postwar days' concern over nuclear benefits and hazards: whether radiation would turn out to have even more bizarre biological effects than had so far been discovered, or whether the first atomic detonation would consume the oceans.

The group thought that in general society approaches such questions by asking, first, what the empirical facts are about the situation, and then moving on to judge those facts in a much broader social-values framework. The problem of the first stage is to determine whether the empirical knowledge--that is, the science--is sufficiently broad and accurate. The problem of the second stage is to know what normative approach to take, how to judge the desirability of benefits that might accrue now and in the future and the acceptability of the risks that might be incurred. The third stage is liable to all the usual pitfalls of social policymaking.

Rational assessment of the DNA risks, no matter what particular form it takes, can be seen as proceeding in stages. What organisms are to be employed in the experiment? What fundamental genetic modifications are likely to be made? What are the chances that the experimental subjects will escape from the containment facilities? If they get out, what is the chance that they will survive? What is likely to be their interaction with the biosphere? Are they likely to infect human beings, wheat, birds, or what? If they infect, what is likely to be the consequence? And so on.

Appraisal of any hazard is likely to involve a long series of such questions. Some in our group thought numbers can be put on all these stages, even if the numbers are uncertain. Others insisted that because we really do not know the numbers well enough, the "bottom line" varies over such a wide range that conclusions really just cannot be drawn.

As has been stated throughout the Forum, there is a finite chance that some terrible things might happen, and there is a finite chance of reaping great benefits (in pharmaceuticals, or improved crop strains).

The fundamental questions for now, given the uncertainty of the numbers, is whether it is legitimate to draw the bottom line at all. For the extremes we seem to be ready to. For both the trivial cases and the obviously horrible cases, consensus is easily reached. As usual, the problem is the cases in the middle.

A lesson not only of our working group but of the whole Forum is that there are so many kinds of experiments with so many different kinds of unknowns that it almost misleading to use the term "recombinant DNA research" as though it were a single category of experimentation. The public discussion must mature in this regard.

Important perspective arises when one recognizes that these experiments are now being done all over the world. So the question is not whether we should do DNA research or not. It is rather a question of what kind of research, under what conditions, and with what timing. Experience gained in the United States will have crucial implications for the quality of research done elsewhere.

These experiments have a unique feature, in that appraisal of the risk may of itself carry risk to the public; such is not usually true with other areas of risk assessment. Another special feature is that the effects may be irreversible. Once a new gene is out in the world, it may be with us for a long time.

The working group urges the scientific community and the larger public to devote conscious attention to developing the research agenda. Experiments are not all equivalent. Asking certain questions early along will provide insights that will improve our confidence in later research.

REPORT OF WORKSHOP NO. 3
Public Participation in Decision Making Regarding Recombinant
DNA Research

*Chairman:* Stanley B. Jones
  *Staff Director, U.S. Senate Health Subcommittee*

The first consensus reached by the workshop participants in dealing with
the way laymen should be involved in decisions concerning recombinant
DNA research was dissatisfaction with the notion of "laymen." The basis
for the dissatisfaction was the feeling that, in this case, everyone
except molecular biologists actively engaged in research should be considered a layman. Given this conclusion, the proper question for the
workshop seemed not how to involve laymen, but who should be involved,
and how, in decisions concerning conduct of research with recombinant DNA.

Two major points of consensus were reached by the group in answer to
this question. The first was that wider public involvement is necessary
and important, for several reasons. First, there are moral values at
stake in this issue, and individuals are equally expert with respect to
their own value systems. The varied public outside the molecular biologist community has valid value perspectives that could be helpful if
brought to bear on these issues. Second, and equally important, the
varied public has a right to be involved in such decisions, for several
reasons. Public funds are being spent for this research; and the public,
being the source of the money, has a right to say what is done with it.
In addition, since public safety and public welfare are at stake, the
public has a right to protect its interests.

The workshop participants discussed the difficulty of effectively
monitoring laboratory accidents to detect escaped organisms or damage
of any kind, and some concern was expressed about proceeding before good
monitoring mechanisms are perfected.

Having reached consensus that there should be wider public involvement,
the next question was how wide. Several opinions were offered covering
a broad spectrum of participation. At one extreme, the opinion was that
everyone should be involved--that this issue needs to be carried to the
American people via referenda conducted across the country. Others felt
that some mechanism such as the Commission for the Protection of Human
Subjects of Biomedical and Behavioral Research, with its majority of lay
participants, could represent the public. There was some discussion of
the problems of involving the public and the inevitable 80 percent who
never really get interested or educated to the degree they can offer a
real opinion. This problem leads to a possible loss of opportunities
for progress because that 80 percent tends to come out conservative.
The kind of public participation which surfaced over this issue in
Cambridge, Massachusetts, was discussed and generally regarded as constructive.

There was discussion of the scientists' role in this wider public involvement. In particular, there were expressions of enormous frustration

at how dependent the lay public is on experts in this area. For example, even after hours of attempts to understand, and the many months that the citizens on the Cambridge Review Board put into this subject, understanding still eluded them. In the end, we have to come back to the molecular biologists for opinion. This led to talk about the responsibility of the scientific community to take the initiative and offer the information the public needs. There was a plea for scientists not to give up, and not to tire of the debate, even though the time lag between consciousness of the issue in the scientific community and awareness of it in the general public is apt to be long.

There was talk of scientists' responsibility to make the key distinctions that the public needs to have made in order to make good decisions. One example offered was distinctions between research in university laboratories and research in the private sector. And, there was acknowledgment that scientists have varied tremendously in terms of their involvement in the debate and their readiness to work with the public, as well as in terms of the issue itself. This variation in style and viewpoints among scientists makes it doubly difficult for the public, but again the plea was "don't tire or withdraw from the debate."

The second major point of consensus reached during the workshop was the need to improve the mechanisms for allowing wider public involvement in decisions concerning DNA research. Here again, several suggestions were offered including the previously mentioned public referenda and the National Commission for the Protection of Human Subjects model. The latter was objected to by several people in that the lay participants on the Commission are not responsible to the public; they are only responsible to themselves. Also suggested were: the model of the Cambridge-wide Biohazard Committee appointed by public officials; health planning agencies with their boards of 51 percent consumers; professional societies like the AAAS; and, finally, Congress as the ultimate representatives of the people. The problems that Congress has with highly charged moral issues in which there is widespread disagreement was noted, however, and there was strong sentiment voiced that congressional involvement should not preclude local community involvement and action. The public does have a record of finding varied mechanisms for impacting on matters like the SST and environmental concerns--ranging from local, grass-roots campaigns to influencing Congress and the executive branch.

The tone of the workshop seemed to the chairman to raise the basic question, what institutions can we trust to speak for the people? The value issues involved in recombinant DNA research are complex, and choices are not clear cut. Many participants in the workshop expressed fear that public values might not be well served or well expressed by the scientific or university communities acting alone. There was also a tone of mistrust of government institutions. In fact, for some present, there seemed to be a real question of whether there is any clearly trustworthy institution in our society that can represent the public interest. In fact, the more distrust of institutions expressed by an individual, the more he/she wanted individuals to be involved in decision making--even if DNA research had to be postponed for ten years in order to get all 215 million Americans involved. At the other extreme were some participants anxious to let science go, and see what they would come up with. Finally, in the middle, were the majority of participants groping for the right institution to represent them and assure that their interests would be heard and considered.

REPORT OF WORKSHOP NO. 4
National and International Efforts to Develop Guidelines:
Should There Be Voluntary or Enforced Rules of Conduct?

*Cochairmen:* Charles Weissman
*Professor of Molecular Biology and Director, Institute for Molecular Biology, University of Zurich*

Harold P. Green
*Professor of Law, National Law Center, The George Washington University*

*Rapporteur:* Elizabeth L. Clark
*Government Contracts Specialist, Upjohn Company*

There was general agreement that, at least for a time, there is a necessity for restrictions on recombinant DNA molecule research. In addition, it was the consensus of the workshop that the restrictions should be flexible in nature so that they might be removed, strengthened, or relaxed in the light of new knowledge. In particular, concern was expressed that legislative measures, as opposed to guidelines, might be unduly inflexible. On the other hand, it was recognized that guidelines might not be universally applicable or sufficiently enforceable.

The uncertainties inherent in the present state of scientific knowledge were emphasized. There was considerable sentiment in favor of measures to require or to promote disclosure to government authorities of plans for prospective research and the results of research conducted.

The workshop heard reports on the measures adopted in Western Europe, the United Kingdom, Canada, Soviet Union, Australia,* and New Zealand. Although most of these countries seem to follow the NIH guidelines, some have adopted some features of the United Kingdom system, at least to the extent that these are more stringent. Several foreign scientists expressed their doubt that their countries would adopt legislation to regulate this research in the foreseeable future.

*Report from Professor Gordon Ada follows.

It was generally accepted that common standards should be adopted on a worldwide basis, but doubt was expressed that this could be accomplished within a reasonably short time scale.

All members of the workshop placed a high value on scientific inquiry and were of the view that freedom of inquiry should be limited only to protect against actual hazards. There was, however, no agreement as to the extent of such hazard at the present time or where the burden of proof should rest as to the presence or absence of hazard.

It should be observed that all participants in the workshop believed that research to increase the fund of human knowledge is an objective of paramount importance. It was recognized that other persons do not regard scientific research as necessarily being an unmitigated good, but no one present articulated this view.


RECOMBINANT DNA MOLECULE EXPERIMENTATION IN AUSTRALIA*

In 1974, when a group of distinguished American scientists called for a voluntary moratorium on certain types of experiments involving the combination of DNA molecules from different sources, the Council of the Australian Academy of Science appointed an ad hoc committee to alert Australian scientists to this situation and to ascertain the extent to which scientists in this country might wish to carry out experiments of this nature. Two of the committee members, Dr. J. Peacock and Professor J. Pittard, subsequently attended the international meeting held in the United States in February 1975.

On the return of Peacock and Pittard to Australia, the ad hoc committee recommended to the Academy Council that a permanent committee be established (the Committee on Recombinant DNA Molecules), with the following terms of reference:

1. Establish a set of guidelines for both physical and biological containment procedures appropriate to the level of risk involved in particular classes of experiment, and if necessary arrange training of personnel in laboratory safety.
2. Review research proposals dealing with the *in vitro* production of novel recombinant DNA molecules and recommend conditions under which these experiments may be carried out. It is recognized that the committee may recommend that some experiments should not be carried out.
3. Collect and disseminate information in this field of research.
4. Liaise with national committees of other countries and with any international organization.


* By G. L. Ada, Chairman, Committee on Recombinant DNA Molecules. Australian Academy of Science.

## Present Australian Situation

The committee was formed and has eight members, about half of whom could be considered as having particular expertise in the area. It should be noted that the Australian Academy of Science, though receiving government funds, is an independent body. It neither controls any research activities nor disburses research funds within Australia. This aspect was an important factor in the council's decision to accept the obligation of administering this service for Australian scientists. The first task was to advise all universities, research institutes, government departments, grant-giving bodies, etc., of the Academy's intention; all expressed approval and agreed to cooperate. Furthermore, committee members have also assiduously explained the work and role of the Academy committee at meetings of appropriate scientific societies throughout the country. The committee published an abbreviated set of guidelines (*Search* 7, 1976, p. 12) but, in practice, both the NIH guidelines and the United Kingdom's Williams report are extensively consulted; overseas advice may be sought where appropriate.

Australian scientists wishing to work in this area are asked to contact the Academy office either directly or via a grant-giving body. They are sent a copy of the guidelines and a questionnaire. Applicants applying for funds from the major grant-giving bodies must also state in their application whether the proposed work is in this area. One or more members of the committee interview the scientist(s) concerned and inspect the laboratory facilities. A report is written outlining safe conditions for the performance of the experiments and submitted to the full committee. When agreed to, these conditions in the form of *Recommendations* for the safe conduct of the experiments concerned are transmitted directly to the applicants or to the grant-giving bodies.

The Academy Committee is fortunate in one respect. Australia has a high-quality but small scientific community with expertise in this area. Members of the committee believe they can nominate those laboratories in this country where this type of work might be done. The committee relies on peer pressure; only recommendations are made. The cooperation of Australian scientists has been excellent. Some seventeen applications have been dealt with in the first year of operation. So far, only minimum- and low-risk experiments have been approved; negotiations are in progress to have the use of at least one facility suitable for moderate-risk experiments.

## International Aspects

Biological materials entering Australia from abroad by conventional channels must be approved by quarantine officials of the Federal Department of Health. The latter now seek the opinion of the committee regarding entry of, for example, hybrid plasmids. The committee does not attempt to impose any restrictions on Australian scientists who wish to carry out this type of work in overseas laboratories other than to advise them that quarantine permission must be obtained on any biological material sent to or brought into Australia.

Maintaining surveillance programs of this type is expensive in time, money, and effort. We would hope that under the auspices of this meeting, either national or international authorities will be encouraged to instigate specific experiments to evaluate strictly, for example, the dangers of specific shotgun experiments. It is important to establish or refute the hazards of experiments of this nature.

Part of our brief is to collect and disseminate information in this area. We are particularly interested in the proposal to establish in the United States banks of DNA segments from various sources. We would hope that the Australian committee could have access to such material on behalf of Australian scientists.

On behalf of my committee and the Australian Academy of Science, I would like to take this opportunity to warmly thank the NIH for keeping us abreast of events in this area. This service has been of great help to us and we would wish it to continue.

REPORT OF WORKSHOP NO. 5
What Is the Appropriate Role of Sponsoring Institutions and of Federal, State, and Local Governments in Relation to Recombinant DNA Research?

*Chairman:* Clifford Grobstein
*Professor of Biology and Vice Chancellor for University Relations, University of California, San Diego*

The workshop reviewed information available on activities at four levels: institution, local community, state, and federal. At the institutional level it heard reports from Texas Tech, University of California at San Diego, Colorado State, Princeton University, Indiana University, and University of California at Irvine. There is variation in the composition of biohazards committees, with some institutions having a majority of biological scientists and a minority of other disciplines and lay persons while other institutions have a majority of nonbiologists and lay members. The workshop consensus appeared to favor strong, non-biological representation. It also favored close liaison with community committees when such are established. In one or two instances, smaller institutions reported difficulty in establishing an effective institutional committee. The possibility of a regional arrangement was discussed.

Reports on community action were received from Cambridge, San Diego, and Princeton. The first two had community committees study the issues and file reports. The reports acknowledged or approved the P3 localities proposed at the local institutions but advocated greater stringency of control than provided in the NIH guidelines.  The issue was raised also of the requirement for compliance by universities to local ordinance. It was clear that this probably varies from place to place.

Reports were received of legislative consideration or action in New York, California, Michigan, and Massachusetts.  Only in the first two has actual legislation been drafted or introduced (in the last week). The bills differ in details but establish state jurisdiction for bio-hazardous activities.  Both bills involve licensure.  Hearings are expected in the next month or so.

At the federal level, Senator Bumpers (S.621) and Congressman Ottinger (HR.3191) have introduced identical bills.  The House version has been referred to the Rodgers Committee, which reportedly will begin hearings March 17, 1977.  Hearings in the Senate are forecast for mid-April.  Congressman Solarz of New York has introduced HR.4232, which would establish a commission to consider genetic research.

The workshop developed a strong consensus that recombinant DNA research can be expected to continue and that federal legislation is now desirable that will:  (1) provide a statutory base for regulation; (2) extend coverage to all research sectors including industry; and (3) preserve flexibility to adjust to increasing knowledge and changing circumstances. The legislation should emphasize maximal community participation as is appropriate to each local area.


A PROPOSED POLICY PROGRAM FOR RECOMBINANT DNA RESEARCH*

The premise for immediate legislation is that the current uncertainty of risk-benefit assessment is high enough to require:  (1) deliberate systematic and effective surveillance; (2) planned research aimed at reducing the assessment uncertainty; and (3) a comprehensive restudy to provide a base for definitive policy.  Accordingly, the following steps are required:

1.  Federal legislation should provide a provisional statutory foundation to underpin and extend the area of application of the existing regulation under the NIH guidelines.

2.  The guidelines should apply to *all* research and applications involving recombinant DNA techniques.  Suitable mechanisms should be added to ensure equal regulation and compliance in federally and non-federally funded sectors.

3.  Flexibility of the guidelines should be retained so as to allow surveillance and regulation to be adjusted to fit developing knowledge and changing circumstances.

4.  Sufficient authorization and add-on resources should be provided to support effective function of institutional biohazards committees


*Personal Comment:  Clifford Grobstein.

or their equivalent, as well as for a program of regulatory research
to improve risk-benefit assessment.

5. An optimum mechanism would be an interim joint executive-legislative
commission to oversee operations and to undertake a new and comprehensive
assessment of all issues. While emphasizing the importance of individual,
community, and ecologic safety, the commission should also carefully consider such broader, long-range matters as priorities to be assigned to
potential applications, ethical and social considerations, the possible
inadvertent or planned impact on evolution of the ecosystem, and the effects of regulation on scientific progress.

6. The appraisal by the federal commission should precede and provide advice to all contemplated local and state actions.

7. The federal interim commission should present recommendations
for additional legislation in two years and should be supplanted in
three years by such definitive provisions and mechanisms as may by then
seem necessary.


REPORT OF WORKSHOP NO. 6
The Use of Recombinant DNA Research in Biological Warfare is
Ruled Out for Nations by the Biological Weapons Convention
of 1972 and the Geneva Protocol of 1925:  Is It Conceivable
That Terrorists or Nations May Try To Use This Technology
To Develop Weapons?

*Chairman:*  Matthew S. Meselson
        *Thomas Dudley Cabot Professor of the Natural Sciences,*
        *Chairman, Department of Biochemistry and Molecular Biology,*
        *Harvard University*

*Rapporteur:* James M. McCullough
        *Senior Specialist in the Life Sciences, Congressional*
        *Research Service, Library of Congress*


The chairman of the workshop opened the informal discussion by suggesting that specific limiting objectives be selected in order to focus
the discussion. As generally agreed upon by the group, these topics
were identified as follows:

1. Does the BW Convention prohibit the use of DNA recombinant
techniques for the development of biological warfare agents?
2. Does the U.S. unilateral renunciation of biological weapons include weapons involving the use of DNA recombinant techniques?
3. What other domestic or local constraints are there against the
use of DNA recombinant research for the development of biological
weapons?

4.  Is there any likelihood that DNA recombinant research offers a threat from the hands of terrorists?

The group discussed the Biological Weapons Convention for some time and in general determined that, within the limits of the information available, the convention contains within its negotiating history and in its language a prohibition making it illegal to use DNA recombinant techniques for the development or production of biological warfare agents and/or weapons.  This view has been explicitly stated by the governments of the United Kingdom* and the United States.  In particular, it was stated by U.S. government arms control and legal specialists present at the workshop that the scope of the convention extends to all forms of recombinant DNA having an action on living cells regardless of the nature of the DNA or the mode of its production.  The same broad prohibition was considered to be embodied in the U.S. unilateral actions in 1969/70 renouncing biological weapons.

Some members of the group questioned the effectiveness of the treaty in the case of a nation that strongly desired to use DNA recombinant techniques for military purposes.  Here, however, the issue is the same as it is with any law.  Making an act illegal does not in itself provide absolute assurance that the act will not be performed.  It does, however, produce social and political pressures through the mechanisms of law and customary practice over time to reduce the probability of an act.  It was also pointed out that, limited as it might be, the treaty does have a provision for complaint and investigation should there be suspicion that the treaty has been violated.

A discussion then ensued as to the applicability of the BW Convention to DNA recombinant research in domestic law.  A member of the group pointed out that at least one legal position was that, under U.S. constitutional processes, it is necessary that separate legislation be enacted to make the provisions of an international treaty domestic law. Article IV of the convention specifically notes that the signatories would take such action.  Upon further discussion it was determined that the U.S. Congress had indeed received legislation proposed by the

---

*Statement from Edward C. Glover, Second Secretary, British Embassy, Washington, D.C.:

In the view of Her Majesty's Government the 1972 Biological Weapons Convention's prohibition of the production of any microbial or other biological agents of types and in quantities that have no justification for prophylactic, protective or any peaceful purpose includes those which might be reproduced by recombinant DNA techniques.  It should also be noted that the UK Biological Weapons Act of 1974 makes it an offence for any person or group to develop, produce, stockpile or retain any biological warfare agents for hostile purposes.

executive branch, which would have fulfilled the requirements of Article IV, but that this bill had never been enacted.

In the absence of more specific data, the members of the discussion group agreed that it would serve a useful purpose to take note of this point and report to the Forum the possible need for further consideration of such legislation.

Two other issues occupied most of the remaining discussion of the workshop. These were the potential danger from terrorist actions involving the direct or implied threat from DNA recombinant techniques and the potential for sabotage of a laboratory in which relatively high-risk DNA recombinant research is going on. The latter point was not discussed in great detail.

With regard to the potential value to terrorists of DNA recombinant techniques, the discussion brought out a number of points. On the one hand, it was recognized that the potential for creation of some molecule or organism that would be toxic or otherwise hazardous could be present. On the other hand, there was a strong impression that so many weapons are already available that there would be little for terrorists to gain from attempting to create a highly problematic new weapon. For example, guns and explosives offer ready and predictable types of threat. If a biological threat was perceived by terrorists as offering some unique advantage, then existing organisms present in nature already offer an opportunity for terrorists to engage in such threats without turning to DNA recombinant techniques. Moreover, the view was expressed that, in spite of impressions to the contrary, there exists no sure prescription for producing highly pathogenic molecules or organisms by recombinant DNA techniques. It was also noted, however, that the psychological aspects of the DNA recombinant technique might appeal to the illogical perceptions of a terrorist group, even if no effective weapons could be produced.

There appeared to be no disagreement with the view of the group that public scrutiny and discussion of all DNA recombinant research was one of the best long-term measures to prevent antisocial uses of this new technology.

The workshop conclusions were summarized as follows:

1. The workshop took note of the fact that any use of recombinant DNA technology for military purposes is prohibited by the Biological Weapons Convention of 1972 to which the United States is a party.

2. The workshop also took note of the fact that in 1973 the administration submitted to the Congress a bill to have the provisions of the Biological Weapons Convention implemented as domestic law. No legislative action was taken.

3. The general view of the workshop was that recombinant DNA technology does not offer any ready means for the development of weapons by terrorists. The view was also expressed, however, that terrorists might attempt to create fear by threats involving the use of such technology.

4. The workshop endorsed the principle of openness and freedom of information to avert the misapplication of recombinant DNA technology.

## REPORT OF WORKSHOP NO. 7
## Effectiveness of Physical and Biological Containment

*Chairman:*   Roy Curtiss III
*Professor of Microbiology, University of Alabama, Birmingham*

*Rapporteur:*   Josephine Clark-Curtiss
*Research Associate, Department of Microbiology, University of Alabama, Birmingham*

There were approximately forty attendants at the workshop that considered the description, effectiveness, and potential improvement of physical and biological containment for recombinant DNA research. Physical containment facilities and procedures are designed to reduce the probability of escape of recombinant DNA materials. Emmett Barkley described the most important features and differences for the P1, P2, P3, and P4 levels of physical containment.[1] In response to a query as to availability of data on effectiveness of different levels of physical containment, it was pointed out that Dimmick *et al.*[2] have published data on inhalation doses for different laboratory procedures that generate aerosols. Other data[3] indicate that the inhalation doses decrease about 100,000-fold when the procedure is done in a class I or II biohazard hood compared to conducting the operation on the open lab bench, and another 10,000-fold if conducted in a sealed class III glove box. The inhalation doses are 100 to 1,000 times higher for the individual conducting the operation than for another individual in the same lab. There were numerous questions about correct use, testing, and failure of biological safety cabinets. Tests need to be performed anytime a hood is moved and on an annual basis thereafter to determine that air flow rates are adequate and that there are no leaks in the HEPA filter. It was also pointed out that the effectiveness of safety hoods could be compromised by improper placement of items in the hood and by rapid changes in air current in the lab as might occur by opening the door, etc. In terms of failure of air exhausts from hoods and P3 labs, it was mentioned that inexpensive air flow meters with warning lights are available.

Concern was expressed about power failures and the need for emergency power generators. Although there was no consensus, the idea was expressed that if the experiment was of such risk that a power failure would be hazardous then the experiment should be conducted in a P4 facility. It was also pointed out that labs should be equipped with emergency lights in the event of a power failure and that investigators should immediately vacate the facility. The question of sterilization of air exhausted from P3 labs was also raised. It was pointed out that this would not appreciably reduce escape of recombinant DNA materials from the laboratory. This is because appreciably more materials escape through the open fronts of biological safety cabinets, by getting on investigators who then carry some of these materials outside the lab, than are exhausted from the facility.

A question was asked about autoclaves in the P3 facility versus in the same building. Either are permissible in P3, but materials to be autoclaved elsewhere in the building require disinfection and then transportation to the autoclave in closed leakproof containers. This procedure, however, was acknowledged to allow for greater human error.

The issue of insect and rodent control was extensively discussed with numerous good ideas. The importance of the integrity of walls and floors was stressed as was the need for cleanliness and good housekeeping. The resistance of cockroaches to and the biohazards associated with extensive use of insecticides were discussed, with no consensus as to the best control measures. The idea of placing equipment in pans containing mineral oil was suggested as a means to prevent insects from getting to recombinant DNA materials.

In terms of facility location, it was pointed out that certain upgrading of P3 facilities had been done in urban environments and in buildings containing patients, especially those compromised by disease. This was deemed a good idea. It was agreed, however, that it was best to locate recombinant DNA research facilities as far away from patient care areas as possible.

With regard to assessment of adequacy of facilities and practices, it was suggested that nonscientists might see problems overlooked by scientists. Thus such individuals should be members of local biohazard committees. Inspection and certification of facilities is the responsibility of institutional biohazard committees except for P4 facilities, which must be inspected by NIH.[1] Facilities in industry were described and discussed and seem to be more ample in space and better equipped than those in academic institutions.

The idea was expressed by several that overkill in terms of physical containment could increase hazards by breeding contempt and thus noncompliance with acceptable practices. It is the principal investigator's responsibility to ensure that all lab workers and others present in facilities engaged in recombinant DNA research follow the appropriate practices. It was generally agreed that this might be easier to accomplish in industrial labs than in university labs.

The idea for a repository for P3 lab plans, facility use manuals, and accident plans was put forth as a means to allow investigators and/or institutions to learn from each other. The question of whether and how to isolate an individual exposed to a recombinant DNA accident was raised but not discussed.

Biological containment is the use of genetically altered hosts and vectors to reduce the probability of survival of recombinant DNA should the organisms or vectors escape from the physical containment facility. The EK1, EK2, and EK3 designations for *E. coli* host-vectors as given in the NIH guidelines[1] were described and discussed. Phage vector components of EK2 host-vector systems generally have mutations to greatly reduce the likelihood of lysogenization and plasmid formation and that cause its replication to be dependent on the host. The host may also possess mutations to diminish survival of any lysogens or plasmids formed. In plasmid-host systems, containment is principally dependent on the host, although modifications of plasmid vectors to reduce the probability of their transmission has been achieved. Disarmed hosts, such as χ-1776,[4] possess constellations of mutations that prevent DNA and cell wall syntheses, which cause death if the escaped cells attempt to grow; another constellation of mutations that confer increased sensitivity to environmentally encountered substances and agents such as bile, detergents, chemical pollutants, and sunlight; and another constellation of mutations to reduce the likelihood that recombinant DNA could be transmitted to other robust microorganisms encountered in nature.

Questions were asked about available data on survival of recombinant DNA in various EK2 systems. This was described and discussed since much information is not yet published. For plasmid vector-host systems, χ-1776 cannot survive passage through the rat intestinal tract and, because of its sensitivity to bile, may be killed in the small intestine. χ-1776 is extremely sensitive to sunlight and is effectively killed by detergents and other chemical pollutants likely to be encountered in wastewater, sewage, rivers, etc. After drying or when suspended in pure water, χ-1776 does not die at appreciable rates but in these metabolically inactive states is unable to transmit recombinant DNA at detectable frequencies. In terms of conjugational transmission of non-conjugative plasmid vectors, available data suggest that this would be more likely at 37°C (i.e., in the intestine) than at lower temperatures in the environment since most conjugative plasmids found in enteric bacteria are unable to engage in conjugation below 27°C. Although χ-1776 is resistant to most known transducing phages, it might be sensitive to unknown phages. Knowledge on quantitative aspects of *E. coli* phage ecology is lacking, although present plasmid cloning vectors are much less well transferred by known transducing phages than are chromosomal genes. The question of the fate of DNA released by disabled hosts lysing in the intestine was raised. The rat intestine contains high concentrations of nucleases that very rapidly degrade DNA,[5] thus further diminishing the chance that recombinant DNA might escape by transforming resident intestinal flora. The only reservation is whether the intestines of recombinant DNA researchers are like those of rats. In terms of available EK2 lambda phage vector-host systems, the Charon phages[6] are unable to lysogenize the host or form plasmids at detectable frequencies, thus killing all infected cells. These phages are propagated on a partially disabled host that is unable to synthesize DNA and its cell wall outside of the laboratory. Since free lambda phage is sensitive to various environmental conditions and since suitable sensitive host strains that could propagate Charon phages are extremely rare in nature, recombinant

DNA contained in these phage vectors is unlikely to survive and be per- petuated in nature.  More extensive tests on components of EK2 systems are now being performed by six NIH contractors.

In terms of improvements in biological containment, several ideas were expressed.  These included use of plasmid vectors that are dependent on the host for replication, the use of nondrug resistance markers for selection, and the use of control signals on the vector to cause cloned DNA to only be expressed under laboratory-controlled conditions.

Frank Young described certain potentially advantageous features of *B. subtilis* as a safer cloning system.  Particularly intriguing is the availability of a phage vector that can only replicate at 48°C.  *B. subtilis* is seldom an animal pathogen and is not a plant pathogen, and no one in attendance at the workshop was aware of any plant pathogens with which it might exchange genetic information.  The sporulative ability of *B. subtilis* can be eliminated by appropriate mutation, and such mu- tants probably survive less well in soil than *E. coli*.  It was agreed that the system warrants further development and study.

In terms of general issues related to physical and biological contain- ment, there was expression of doubt on the accuracy of estimated proba- bilities for escape and/or survival of recombinant DNA materials and also concern that biological containment not be substituted for physical con- tainment as is now permitted by the NIH guidelines.[1]  It was agreed, however, that the potential for human error may be the weak link in achieving the levels of containment afforded by available physical and biological containment facilities and systems.  Because of this, it was deemed advisable to not overcrowd physical containment facilities and to not conduct research with more robust hosts and/or vectors in a facility using disarmed host-vector systems.

REFERENCES

1.  NIH guidelines for research involving recombinant DNA molecules. Fed. Regist. 41:27902-27943, July 7, 1976.
2.  Dimmick, R. L., W. F. Vogl, and M. A. Chatigny. 1973.  Potential for accidental microbial aerosol transmission in the biological laboratory, pp. 246-266.  *In* A. Hellman, M. N. Oxman, and R. Pollack (eds), *Biohazards in Biological Research*.  Cold Spring Harbor Laboratory, Cold Spring Harbor, N.Y.
3.  Curtiss, R., III. 1976.  Genetic manipulation of microorganisms: potential benefits and biohazards. Annu. Rev. Microbiol. 30:507-533.
4.  Curtiss, R., III, D. A. Pereira, J. C. Hsu, S. C. Hull, J. E. Clark, L. J. Maturin, Sr., R. Goldschmidt, R. Moody, M. Inoue, and L. Alexander. 1977.  Biological containment:  the subordination of *Escherichia coli* K-12.  *In* Proceedings of the Tenth Miles Internation- al Symposium.  In press.
5.  Maturin, L., Sr., and R. Curtiss III. 1977.  Degradation of DNA by nucleases in the intestinal tract of rats.  Science.  In press.

6.  Williams, B. G., A. E. Blechl, K. Denniston-Thompson, H. E. Faber,
    L-A. Furlong, D. J. Grunwald, D. O. Kiefer, D. D. Moore, J. W.
    Schumm, E. L. Sheldon, O. Smithies, and F. R. Blattner. 1977.
    Charon phages:  safer derivatives of phage lambda for DNA cloning.
    Science.  In press.

REPORT OF WORKSHOP NO. 8
Can the Results of Basic Research with Recombinant DNA Be
Transferred to Industrial Applications?

*Chairman:*     A. M. Chakrabarty
                *Staff Microbiologist, Physical Chemistry Laboratory,*
                *General Electric Research & Development Center*

What are the Processes and Problems Involved in the Application for Patents Involving Recombinant DNA Research?

*Chairperson:*  Betsy Ancker-Johnson
                *Assistant Secretary of Commerce for Science and*
                *Technology*

The chairman opened the workshop and expressed his opinion regarding the
desirability of developing an alternative host system other than *E. coli*
for use in industrial applications.

   Addressing the processes and problems involved in the application for
patents involving recombinant DNA research was Betsy Ancker-Johnson,
Assistant Secretary of Commerce for Science and Technology.  In her con-
cise presentation and in her response to subsequent questions, Dr. Ancker-
Johnson clarified several misconceptions regarding this matter, especially
those concerned with the Commerce Department's order of January 13, 1977,
permitting the accelerated processing of recombinant DNA patent applications.  She stressed that the order did not "exempt" private-sector researchers from compliance with the NIH guidelines, as has mistakenly been
reported, but was designed to gain the adherence of nongovernmentally
funded researchers, both foreign and domestic, to the NIH guidelines.
She explained that requests for accelerated processing must be accompanied
by a statement of compliance with the containment and other substantive
portions of the NIH guidelines.  However, researchers are not obliged to
aver compliance with those procedural portions of the guidelines that, if
voluntarily followed, would occasion the loss of proprietary or patent
rights.  She stressed that total compliance was waived in such cases since
investigators would almost certainly forego "special" processing rather
than risk such losses.  The result, of course, is that no incentive would
exist for compliance with the safety features of the guidelines.  Dr.
Ancker-Johnson further pointed out that accelerated processing was

suspended as of February 24 for all recombinant DNA patents except those dealing with safety inventions.

Apart from this, Dr. Ancker-Johnson focused on several other questions. One such question raised dealt with the nature of inventions for which patents might legally be granted. She pointed out that, except in certain cases involving asexually reproduced, nontuber-propagating plants, living organisms, whether recombinant or not, are *not* themselves patentable. Notwithstanding the fact that a particular organism may not, therefore, be patentable, the process by which the organism is made, or the product resulting from the use of the organism, can be patented. She further stated that the accelerated processing option was limited to recombinant DNA inventions; it does not apply to inventions involving other genetic processes and procedures, such as transduction or transformation. Finally, she indicated the need for a "statute" protecting the confidentially of information disclosed to the government, so as to enable private-sector researchers to comply fully with the guidelines without at the same time having to forfeit valuable property rights.

Several people from other industries presented their viewpoints about those recombinant DNA experiments that pose more imminent or grave hazards in comparison to others that may pose little or no conceivable hazard and may in fact be industrially more attractive. Ronald Cape, President of Cetus Corporation, expressed his belief that the current public discussion of the kinds of controls appropriate to recombinant DNA research signals to responsible executives that prudence rather than thoughtless activity are indicated until the issues are resolved. There are a number of technical hurdles still to be surmounted before the more exciting of the projects frequently discussed can be realized. There are also many pedestrian problems that need to be solved before a laboratory phenomenon can be scaled up to a dependable, economic, and safe process. These points were also emphasized by Ralph Hardy, Associate Research Director of DuPont, and to some extent by Raymond Valentine of the University of California, Davis. Dr. Hardy explained that a number of factors must be favorable for any realistic industrial application of the recombinant DNA technology. These factors include policy for realistic federal guidelines and an adequate useful period of proprietariness, safety to workers, society, plants, animals, and the environment, and economic incentive provided by a new product, for which there is adequate need, or an improved process for an existing product. Dr. Hardy also emphasized the distinction between applications involving use of product of organism versus direct use of organism, that is production of antibiotic, hormones, human interferons, etc., by controlled fermentation with engineered microorganisms in highly monitored microenvironments, as against the use of intact microorganisms such as $N_2$-fixing engineered strain, $N_2$-fixing crops, or photosynthetically efficient crops in a macroenvironment.

These viewpoints were further elaborated by Irving S. Johnson, Vice President of Lilly Research Laboratories, who presented his views about the nature of regulations that might be appropriate for monitoring the recombinant DNA research in pharmaceutical industries. There was considerable discussion on whether there should be local regulations as well as state and federal regulations, the form of regulations if any, and the enforcement of such regulations. The salient outcome of this discussion was as follows:

1. The recombinant DNA research may provide significant benefits for the public health, agriculture, and the environment.
2. This type of research may also carry attendant risks. In this regard, it should be noted that the application of the National Institutes of Health guidelines should limit these risks. Incidentally, the pharmaceutical industry has in large measure supported the guidelines and made recommendations for their improvement.
3. Any controls applied can have a detrimental effect on innovation and the development of new knowledge and its dissemination. Freedom of scientific inquiry is a very precious thing, and efforts to limit the pursuit of new knowledge can be very harmful.

In the final viewpoint, federal legislation should regulate and establish standards for the physical facilities where recombinant DNA research is conducted; however, regulation of the actual research carried out in those facilities should not be operative.

In all probability, regulatory authority should fall within the functions of the Secretary of Health, Education, and Welfare. Recognition should be given to the potential contribution of recombinant DNA research in agriculture as well as in the health field.

Whatever statute is drafted should recognize the potential contributions to the health and well-being of citizens afforded by recombinant DNA research, as well as providing a basis for minimizing the potential hazards of this activity.

A law covering recombinant DNA research might include some of the following kinds of thing:

● Facilities engaged in such research should be registered with an appropriate agency within HEW. The registrations which they submit should indicate: the facilities are adequate for the type of research contemplated; the facility has adequately trained personnel; what the facility's level of containment is; and that those involved in research at the facility will follow the guidelines.

● A federal act should provide for inspection so the government can determine if the information in the application is correct. Such inspections would also permit a determination if the facility was in compliance with the guidelines. The federal inspectors might want to meet with the biohazards committee to gain first-hand knowledge on the procedures utilized at the laboratory.

● There will have to be some kind of teeth in any federal act that will permit the Secretary of HEW to stop recombinant DNA research at any specific facility if that laboratory is not following the guidelines or if the procedures at that laboratory are endangering the community or the health of workers. Undoubtedly, there will also have to be

penalties for those that engage in this activity without registering
and following the guidelines.  This can be dealt with by appropriate
legislative bodies.

## REPORT OF WORKSHOP NO. 9
### How Natural Is the Exchange of Genes Between Unrelated Bacteria?--The Benefits and Risks of Prokaryote-Prokaryote Gene Exchange

---

*Cochairmen:*   Stanley N. Cohen
*Professor of Medicine, Stanford University Medical Center*

Herbert W. Boyer
*Professor of Biochemistry, University of California Medical Center, San Francisco*

In the context of the workshop, *natural* was defined as "occurring without
the intervention of recombinant DNA techniques." *Unrelated bacteria* were
defined as bacteria that are members of different taxonomically distinguishable genera. *Prokaryotic organisms* are simple organisms, such as
bacteria, in which the DNA is not contained within a nucleus but instead
is dispersed in the cell.

There were three primary goals of the workshop:

1.  to examine currently available evidence on whether gene exchange
occurs between unrelated bacteria without the intervention of recombinant
DNA techniques;

2.  to discuss the potential benefits and hazards of prokaryote-gene
exchange through intervention by *in vitro* techniques and by other
(physiological) methods of exchange; and

3.  to determine what additional information is needed for more
definitive evaluation of the extent of natural gene exchange among prokaryotic organisms.

In examining the evidence for natural gene exchange among prokaryotes
during the first part of the workshop, every effort was made to maintain
rigorous standards for the use of experimental evidence to document all
statements.  In particular, the workshop participants considered it
important to explicitly distinguish experimentally documented facts from
opinions.  For the second and third parts of the workshop, we jointly
agreed that much of the discussion would necessarily have to deal with
opinion, but we attempted to be explicit about the conceptual basis for
opinions offered.

The cochairmen of the workshop invited a series of scientists who
were known to hold widely divergent points of view about the naturalness

of gene exchange among microorganisms, and about the benefits and risks of prokaryotic gene exchange, to participate in the workshop and to suggest other participants who shared their views.  The individuals contacted included Frank Young, Ronald Olson, Louis Barron, Arnold Demain, Stanley Falkow, Jeffrey Schell, Eugene Nestor, Jonathan King, Gordon Edeland, and Richard Goldstein.  Several of these individuals gave oral presentations at the workshop, whereas others communicated written documents that were read in their entirety or were excerpted by the chairmen during the workshop.

The data presented during the first part of the workshop indicated that gene exchange without the intervention of recombinant DNA techniques is exceedingly common between species and genera that are recognized as being taxonomically distinguishable.  Gene exchange between unrelated bacteria occurs most frequently in nature when the two species occupy a common habitat.  Gene exchange was also noted to occur between microorganisms that inhabit different ecological niches when the organisms are related.

Because exchange can occur between organisms that have either habitat or ancestry in common, extensive opportunities for transfer of genetic material among different prokaryotic species exist in nature.  In particular, compelling evidence was presented to indicate that most or all Gram-negative organisms can exchange genetic information outside of the laboratory, and without the use of *in vitro* recombinant DNA techniques.  However, most studies of this process thus far have been limited to exchange of genes present on extrachromosomal DNA molecules such as plasmids, and it was noted that less information is available about the natural exchange of chromosomal genes between unrelated bacteria.  While many plasmids are known to acquire chromosomal genes from their host bacteria, the possibility remains that some examples of intergeneric gene exchange may be found to be restricted only to plasmid genes.

Within Gram-negative organisms, exchange of genetic information occurs between such taxonomically distinct species as *Myxobacteria, Neisseria, E. coli* and other Enterobacteriacae, the photosynthetic bacterium *Rhodosprillum rubra,* and plant bacteria such as *Rhizobium* and *Agrobacterium.*  Since many participants in this Forum are not microbiologists, I'd like to mention briefly that the Gram stain is a property commonly used in the taxonomic classification of bacteria.  Organisms that can be stained with the Gram stain are classified as "Gram positive," while others are termed "Gram negative."  The ability of the bacterial cell to pick up the dye and be stained by it is thought to be determined by the biochemical nature of the bacterial cell surface.

Another property that is used taxonomically for distinguishing organisms is morphology or shape.  Some bacteria are long and rod-shaped organisms, while others are circular.  Classification can be determined additionally by the presence or absence of particular genes within a bacterial isolate, or by antigenic sites on the cells' surface or on hairlike projections extending from the cell.

In addition to natural gene exchange occurring between unrelated bacteria, there was also evidence presented at the workshop to indicate that the prokaryotic organism *Agrobacterium tumorfaciens* can transfer genetic information contained on a bacterial plasmid to plants that are infected by *Agrobacterium*.

It should be pointed out that conclusions about natural exchange of genetic information among Gram-negative organisms pertain primarily to aerobes and to facultative anaerobes. The exchange of genetic information between obligative anaerobes, that is, those organisms that require the absence of oxygen to grow, has not been well studied.

Evidence was also presented to indicate that transfer of genes occurs among Gram-positive bacteria. Such exchange can involve either related bacteria such as *B. pumulus* to *B. subtilis,* or taxonomically distinct species, such as from *Staphococcus aureus* (a Gram-positive coccus) to *Bacillus,* which is a Gram-positive spore-forming rod-shaped organism. There is also a report in the literature of exchange of genetic information between Gram-negative and Gram-positive organisms, and other laboratories are currently investigating such exchange following the initial report. No information appears to be available at the present time about the extent of natural transfer of genes from Gram-positive to Gram-negative organisms without the intervention of recombinant DNA techniques.

It was also noted by several speakers that genes expressed in one microorganism may not be expressed when they are transferred to another host. Moreover, it was noted that the genetic background of the host to which genes are transferred can affect the stability and persistence of those genes in addition to affecting their expression. Some plasmids that can be transferred intergenerically seem to be maintained in the new host in the absence of continuing active selection, just as they are maintained in the original host, whereas other plasmids that may be introduced by the same nonrecombinant DNA procedures seem to put the new host strain at a biological disadvantage when selection for genetic markers on the plasmid is not carried out continually. Thus, growth of the recipient is inhibited, and those organisms in the population that contain plasmids are gradually diluted out by the more rapid growth of cells that lack plasmids. The extent to which this process occurs seems to vary among different organisms and different plasmids.

Now, on the subject of possible benefits and hazards of prokaryote-gene exchange, this subject was divided into two general categories. Category one concerned experiments involving genes from organisms that naturally exchange genetic information. The second category concerned possible benefits and hazards from prokaryote-gene transfer between organisms that are not known to exchange genetic information without the intervention of recombinant DNA techniques. The conceptual basis for this division was the view that if genes transferred among bacteria using recombinant DNA techniques can also be transferred by natural physiological processes, then recombinant DNA methods do not result in the creation of a novel biotype and do not lead to any novel hazard. In instances where gene exchange can occur naturally, the hazard of the recombinant DNA experiment was deemed to be of the same magnitude as the hazard inherent in working with the same organisms by traditional experimental methods.

For those pairs of microorganisms that do not ordinarily exchange genetic information by physiological processes, it was recognized that recombinant DNA methodology could lead to the construction of a biotype having novel properties and that this theoretically might lead to an alteration of the ecological niche of the recipient organism. However, it was noted by several participants that the capability of an organism to survive in a particular ecological niche is determined by multiple genetic traits, and that recombinant DNA methods commonly used result in the introduction of only a limited number of genes into a single organism. In some instances, the potential for alteration of the ecological niche can be predicted from knowledge of the metabolic capabilities of both donor and recipient organisms.

The potential hazards of experimentation with organisms carrying genes for pathogenicity were discussed briefly. Several speakers expressed the view that the extent of hazards in work with an organism that is the recipient of such genes is unlikely to be of greater hazard than work with the donor pathogen itself.

The benefits of prokaryote-gene exchange were also divided into two categories: research benefits and practical applications. Research benefits noted include the elucidation of the structure, function and evolution of bacteria and bacterial viruses, and an understanding of development of mechanisms of bacterial antibiotic resistance. Applications benefits include the transfer of production genes for synthesis of a variety of gene products made in prokaryotes. These include items such as vitamins, amino acids, antitumor agents, pesticides, animal growth promoters, biologically useful enzymes, and new types of antibiotics. As mentioned earlier, potential hazards from such experiments involve the possible alteration of ecologic niche if genes derived from organisms that do not naturally exchange genetic information are mixed, and also possible hazards of the genes themselves if such genes code for toxic substances. However, it was noted that experiments involving genes that code for toxins are currently not permitted under the NIH guidelines. It was suggested also that recombinant DNA methodology provides a potential method for increasing the yield of currently produced antibiotics, and that experiments to accomplish this do not require interspecies or intergeneric transfer of DNA.

The participants of the workshop felt that the scientists who have been most vocal in their opposition to recombinant DNA experimentation were not sufficiently represented at the workshop to have expressed their points of view about prokaryote-prokaryote gene exchange; such individuals had been invited, as already pointed out. Thus, the workshop group chose not to arrive at a consensus about the relative merits of the benefits and hazards of prokaryote-prokaryote gene exchange in the absence of a forceful presentation from the opponents of recombinant DNA research.

Little time was spent in discussion of the applications benefits. It was noted that there may be special problems involved in the practical application of prokaryote-prokaryote recombinant DNA methodology such as scale-up procedures and procedures for worker protection in commercial laboratories. However, the group did not consider this matter in detail, since it was the subject of another workshop.

The final part of the workshop concerned identification of the information needed in order to understand more fully the extent of natural exchange of genetic information among prokaryotes. First, it was noted that current information about the breadth of gentic exchange among Gram-positive organisms is limited, as reported above. There is also little information about exchange of genes between Gram-positive and Gram-negative organisms and about the extent of exchange between aerobes and anaerobes. These various areas were pointed out as deserving special attention.

It was also felt by the workshop group that, although a great deal is currently known about the natural exchange of genes among Gram-negative organisms, additional information is needed here. In particular, the data presently available concern mostly exchange of plasmid genes; information about the extent of exchange of chromosomal genes is desirable.

Another area worthy of investigation involves the factors that influence the stability and expression of genes in different bacterial hosts. As pointed out by several of the scientists who presented data at the workshop, the biological properties of the recipient organism appear to affect the persistence and expression of genes that are introduced in intergeneric or interspecies transfers.

Finally, it was noted that our current state of ignorance about many aspects of gene exchange in nature is not confined only to the area of prokaryote-prokaryote gene exchange. As noted earlier, there are reports now that *Agrobacterium,* a prokaryote, is able to exchange genetic information with a eukaryotic organism--namely its host plant. There are also several reports in the literature that suggest that genes derived from eukaryotes may be transferred to prokaryotic organisms by bacterial viruses. It appears to be of great importance to obtain additional experimental evidence on the extent and naturalness of these kinds of gene exchange that occur without the intervention of recombinant DNA techniques.

REPORT OF WORKSHOP NO. 10
Ethical and Moral Issues of the Research

*Cochairpersons:*   Ruth Hubbard
                    *Professor of Biology, Harvard University*

                    J. Wesley Robb
                    *Professor of Religion, University of Southern California*

This report is an effort to identify the major ethical and moral issues that were raised in the workshop and some of the value judgments that were made about these issues. Some of the members of the group were concerned primarily with the social and political implications and came to this workshop because they felt it lay closest to their interests.

The chair made a few preliminary statements in order to place the discussion within a larger philosophical context: The ethical question is not essentially a scientific question; rather, it is a philosophical issue. This does not mean that scientific information is irrelevant, but the description of phenomena does not in itself determine the ethical answer. The "ought" is not self-evidently derived from the "is." Such questions as the nature of humankind, the nature of natural processes, and the nature of the human community are scientific, social, political, and philosophical issues. Therefore there is a need for a greater dialogue between the various lay and professional communities in facing the ethical implications of recombinant DNA research.

Some of the following specific ethical issues were raised:

First, do we have the right to meddle or interfere with the order of nature? Some argued that there is a built-in homeostatic process within nature that should be only minimally altered and that great care should be used when we interfere with this process. Uncritical acceptance of science and technology, as applied to the manipulation of nature in the name of progress, has brought us, in part, to the present ecological crisis. The new frontier of recombinant DNA research should be carefully controlled to ensure that it does not move toward a similar disruption of natural processes and bring irreparable harm to us all.

On the other hand, it is difficult morally to defend the notion that there should be no interference with nature--the moral problem is essentially one of control and an evaluation of the kinds of interference that are proposed. Are the consequences irreversible and to what extent are the experimentations actually and potentially harmful to human well-being?

Second, the workshop members generally agreed that the problem of genetic engineering is a relevant issue to the recombinant DNA controversy. These are not issues raised by hysterical neurotics who are motivated by fear. They are very real legitimate moral issues related to the future of humans. Who has the right to make the kinds of decisions that may affect the yet-to-be-born? Are we at the beginning of a new era? Scientists tend to be optimistic about the application of rational and scientific answers to all our problems, but is this optimism justified? The scientific community is a reflection of the mixed motives of the community at large--self-interest, competition, and professional status, as well as humanitarian concerns. Members of the workshop were very concerned that the scientific elite not be left to decide these questions alone, or serve in the principal advisory relationship to government, but that there be full and complete discussion by all parts of the lay community, rich and poor, professional and nonprofessional, religious and secular, concerning the social, political, and ethical issues. Some felt that the research now under way should be done slowly and deliberately; others suggested that it should cease while this discussion is going on.

Human values are at stake and appropriately are the concern of everyone. It was suggested that perhaps the Cambridge City Council experience could provide a model, if the concern had extended beyond the charge by that body regarding research at the P3 and P4 levels. If that model could be extended to include broader issues, it might provide a structure for meaningful discussion below the federal level.

Third, the final ethical issue we discussed concerns the problem of social priorities in relationship to matters of health and nutrition. Some recombinant DNA research now being done applies to this question. However, with the scarcity of medical resources at our disposal, what priority should be given to recombinant DNA research? It seemed to many that the moral question forces us to look more objectively at human health needs than can be realized by those already involved in this type of research.

In conclusion, we had a fruitful and informative discussion of many issues from widely different perspectives, though it was observed with regret that few scientists were present. Could this be due to the fact that many scientists assume that ethical propositions are noncognitive and emotive in nature because they are neither analytic nor empirical statements? We hope not. Deep concern was expressed about the problem that many scientists appear to be uninterested in the philosophical, social, and political framework upon which their scientific world view is predicated.

We would like to assure this body that there were no polemical diatribes; the discussion was serious and provocative. The moral question centers in the problem of preserving human values and at the same time permitting legitimate scientific investigation that might improve our human lot on this planet. It was generally agreed that freedom of inquiry and its application must be within the limits of reasonable and careful social restraints.


## REPORT OF TOWN MEETING

*Chairman:* Ted Howard
*Co-director of the Peoples Business Commission*


Last night at 6:30 we had a small gathering here. It was a town meeting on the question of what is really at stake here: the social, moral, and philosophical implications of recombinant DNA work and research. It was really a symbolic gesture on our part, I think. Many of the people who came here the first night, the protesters, if you will, the people with the banners and signs, the people who stood at the microphones and asked questions, didn't come back. I am sure that that relieved many of you here that they weren't here raising their questions throughout all of yesterday. They probably won't be here today, because you see, for them, they asked their questions. They made their demands, and yet it seemed to all of us that our positions, our viewpoints, our concerns went unanswered. There were many questions asked that were simply not addressed. We were reassured that over the next two days these kinds of subjects, the moral, the social, the economic, the philosophical implications of the work you are engaged in, would be discussed. And yet, I don't believe it was. I think the reason that many of us didn't come back is because we thought the whole thing was a fraud and were frankly not going to participate in it.

So yesterday I spoke on their behalf, the people who didn't come back. There were some people in attendance at the town meeting. I want to go over again what I said last night not only for your benefit, even though you may not want to hear it, but also for the benefit of the press, for some of the people who are here with major publications, such as the *Washington Post* and the *New York Times*, who haven't seen fit to open up to their readership the real controversy here beyond the safety question.

Now maybe my topic last night should have been what is the point of this whole gathering. You know, every speaker, even the people who are the hawks on this issue of recombinant DNA, would come up and say, you know, we want to proceed and there are safety issues here, but it is certainly the most important moral issue of this decade, perhaps that humanity has ever faced. Over and over again this came up, and then everybody proceeded to go forward and talk about the safety issue. It is sort of like a 1984 mentality, very much like we are all kind of zombied out. We all know somehow deep inside ourselves that this is the most important moral question facing humanity, so what do we do? We talk for three days on safety.

The assumption that I believe that you men and women here, the scientific community, have made, is that the moral question has basically been resolved, and the question is how to proceed safely. I would go one step back, and I would say that that moral question has not been resolved. I do not think that we have debated in this society the question of whether this research should proceed at all, whether we are ready in 1977, at the end of the twentieth century, to begin taking the steps along the road that are ultimately going to lead to the alteration of the human species.

I would raise some questions such as these. Should this research be done at all? Who will decide the genetic fate of humanity? Who, here in this room, is qualified? I would never put myself in the position of saying to you that I am qualified to set myself up with some of my colleagues in a committee to decide what is a better human being, what is the perfect human species, where we should go in the evolutionary order.

What is a normal human? What is normal? What criteria are you using? I hear up here that we will be able to turn off the bad genes, that we will be able to turn on the normal ones. We will be able to improve people. Now maybe that might be an easy philosophical question to resolve when we are talking about cancer, but what about fifteen or twenty years down the road? What kind of issue comes up then?

Finally, are we ready to proceed, really posthuman, beyond the human species. Now I know that many of you think this is fairy tale. There was a speaker who stood up here yesterday and basically dismissed these concerns and sort of pooh-poohed them, saying well, what can you expect from a bunch of young people who grew up with science fiction, who watched "Star Trek," and who lived under the nuclear bomb, and of course they are going to come up with these wild scenarios. A few of you are nodding out there now, so I see you are in agreement with that position.

But the point is, ladies and gentlemen, that you need to take that long-range view. You need to take responsibility for looking twenty years into the future, because you are setting the precedents today, there is no way to get around it. Today it is microbes. You see, I think many of you think that we, the lay people, the people who haven't studied science, are dumb, that we don't understand this issue, that we don't have the credentials, the degrees, and so forth. Well, let me tell you we are not so naive and we are not so stupid that we can't look beyond our own noses and beyond our own self-interest to see that what you are doing today with microbes is going to be translated in twenty years into something far beyond what you are experimenting with now.

Maybe you are all people of good will, and maybe none of you intend for this to happen, but the truth is that you have launched this process, and this is the eventual outcome of the work you are involved in.

After I am done speaking, you see, everything will go on just as it is scheduled to go on. We will go into the next topic, food production, and we will talk about new strains of wheat that will feed so many people throughout the world. We will pretend that that is the issue, and we will pretend that we are not talking about the ultimate genetic future of humanity.

So I would say that it is time for this grouping, as the leaders in this field, to take off your scientific blinders and to reflect back ten or fifteen years ago to what happened in this country as the Vietnam war was becoming a concern. There were many of your colleagues in other scientific disciplines who participated in that kind of pure, academic research on campuses around the country that was just for the freedom of their own scientific inquiry and their own interest, their own curiosity. And yet millions of students were very aware of the fact that this pure research argument was in many aspects just a sham, that the pure research being done at Harvard, at Columbia, at the University of Michigan, at the University of California at Berkeley on issues of trajectories and weather modification and so forth, all had serious implications in the war in Vietnam, and those techniques were all applied there.

Now that you are embarked on this course for the future of humanity I would simply say that you should face the facts, realize that this research that you have started is going to be carried to its ultimate conclusion. It is part of the technological society, you know. There is this ethic, this scientific technique ethic that says what can be done should be done and will be done. And just because you are interested in a new strain of wheat, don't be so naive as to assume that we don't know what the ultimate conclusion of this is.

I have gotten my five minutes here, seven minutes, or whatever, you see, and so you will be able to say that there was debate here, that it was all very open, and that it was the most open scientific conference of this kind, there were critics and everything. And you will be able to say that we debated the whole issue, and then you will be able to come around and say nothing changed.

I know we are going to have hearings in the Senate. You are trying to set a precedent there. If there are any people here from Capitol Hill who are going to be participating in those hearings, I want to let you know that this is just the first protest. We are just the little ruffling wind before the storm of public outrage that is going to accompany this issue. And just as there were tens of thousands of people during the Vietnam era who put themselves on the line, who stood up for a principle they believed in, who said to the generals that you don't have all the information, who said to the academic researchers you don't have all the information about the war, we know what is right or wrong. You are going to see that same kind of public controversy on this. And I warn you, if you do not take seriously the social, moral, political, philosophical issues being raised here, there will be new waves of protest, and they are not going to be nearly as gentlemanly I think as what has occurred here. That is not a threat; that is to tell you what is reality. This is the most important social issue of the coming next decade, and people in this country who haven't even heard of this are going to be very concerned when they understand the long-range implications.

Now, my last comment. Last night after a frustrating day here I went home. I turned on the television set like most Americans to try to escape the insanity in this country, and I turned on a very interesting program. It was about the wolf and about what an endangered species it is. I began thinking about the wolves, and the whales and the eagles and the other animals, the tigers that are being driven into extinction in our great technological society, and I realized that ultimately what we are talking about here, and I know that you are not involved in this, you see, because you are all people of good will, but ultimately what we are talking about here in twenty, fifty, perhaps one hundred years, is the final endangerment of the human species as we know it. That is the God's honest truth if you are going to attempt to go beyond humanity. Let me say that unlike the wolves, and the eagles, and the tigers, you see, we are not going to go quietly. We have means at our command to resist the final change in the human species. We will not go gentle into that brave new world, that new order of the ages that is being offered to us here.

# DISCUSSION

HAMBURG: Thank you, Mr. Howard. I think you do underestimate, and seriously underestimate, the extent to which these concerns are today a part of our society, in the Academy, in the universities, in the Congress, in the administration. The transformation which our species has wrought on this planet, particularly since the Industrial

Revolution, this drastic change in our own environment, produced in large part by advances in science and technology, and their eager application embedded in large doses of wishful thinking has created a whole series of dilemmas, in many respects the great dilemmas of human biology, of health, and of disease. And there is not only concern, but I would say a rapidly increasing volume of analysis and research in these matters to try to understand what are the effects we have had inadvertently upon ourselves, and what are the risks of these changes as they accelerate at the present time. The recombinant DNA research is one part of that much broader picture. I don't know whether it is the riskiest. It may turn out to be much less risky than other aspects of the great transformation which we have been undergoing.

I personally would like to see much more effort within the scientific community devoted to these questions, and have so written and published for a number of years. Of course, these considerations, as I emphasized at the outset, go much beyond the scientific community, not only in their effects, but also require going beyond the scientific community for their adequate analysis.

So I think there is some sense in which you are pushing on an open door. But I am glad you make your remarks, although I wasn't enthusiastic about your using the term "fraud," but nevertheless we understand you feel strongly about the issue. You would be surprised how many of us feel strongly about similar issues, too.

In any event, let us now have some general discussion.

ALTHEA AVERY: I am a high-school biology teacher, and I teach in Fairfax County. I have two brief questions. What is the Peoples Business Commission, and where does it get its money?

HAMBURG: Mr. Howard, would you respond?

TED HOWARD: The Peoples Business Commission has been formed for about six months. We have 25,000 members around the country. We are primarily concerned with economic change in this society, challenging corporate power and abuses and influence of giant economic institutions on this government. We put out a lot of educational material, and 5,000 school systems and 3,000 library systems use our material around the country, everything from syllabus and study guides on economics to work guides and so forth.

Over the last five years, we also have been formed as the Peoples Bicentennial Commission. We have sponsored a number of demonstrations and rallies, done everything from writing seven books on economic and historical issues to appearing in testimony before Congress and so forth.

Our money comes primarily from three sources. The bulk of it comes from our 25,000 members who sponsor us to the tune of $10 or $15 contributions a year. As I said, we have written seven books with Bantam, and all the advances, proceeds, royalties from those books go to the Peoples Business Commission. We write articles, lecture at universities and so forth, and we have received some grants from small family foundations. We are a nonprofit educational foundation.

GEORGE WALD: Perhaps I should begin by saying I too am very grateful for the remarks of the last speaker, and in fact as this conference opened I have said before and say here, I think the big issue is a frank and clear and widespread discussion of this entire field, the application specifically of the recombinant DNA technology to the solution of biological problems with all its prospects. I think that question has really not been addressed, and that the last speaker was quite correct. One turns with relief because it is a smaller and more handleable subject to problems of safety. I regret that very much. I wish this larger question could be addressed, and that indeed scientists would take the lead in addressing that question.

Now I too would like to simply comment on some of the things that came up in the workshops. First of all, in regard to Matthew Meselson's report on the possible military applications, it is soothing and unfortunately utterly unreliable, that is, that indeed a large number of nations, 110 in all, including our own, have signed an agreement not to use biological warfare. And indeed the last chairman of the Arms Control and Disarmament Commission has given us an assurance that this convention is applicable to those DNA products also.

But I would like to tell you that a pretty hard-headed group, to wit, the Federation of American Scientists, lead by people who have wide experience in arms control and arms design, take a different view of this situation, and published in April of 1976 a public interest report devoted to recombinant DNA that has this to say, passed and approved by the Council of the Federation of American Scientists:

> Few doubt that this technology has the potential for deliberate misuse to produce great dangers. Genes from disease-causing pathogenic organisms, or from organisms that produce highly toxic agents could be implanted in hosts capable of rapid spread so as to produce dramatic new biological dangers. Not only common sense, but the Biological Weapons Convention of 1972 to which the U.S. and 110 nations have become signatory, demands that scientists eschew development of such agents. Nevertheless, since treaties are neither universal nor self-enforcing, the world must begin to face a biological proliferation threat that might, before long, rival that of nuclear weapons.

One more comment on this, and that is that Meselson used that badly worn and abused term, *terrorists*--we have terrorists to worry about. I want to comment on this. All the big violence that I know of in this world comes through official government actions and is accomplished usually by people in uniform. That is the way it is. And as for this fear of terrorists, please everyone here, worry plenty about what its consequences may be. I have in my possession a big report prepared under contract from the Nuclear Regulatory Commission on the anticipated effects on civil liberties in this country of the necessary precautions to guard against terrorists stealing plutonium 239 from stockpiles, the present stockpiles of nuclear waste products. You should please understand that recombinant DNA as a possible object of terrorist action will get right into the same story. There is a very serious threat to our civil liberties that is being stirred up already in the case of nuclear power. Our public safety will demand serious contraction of our civil liberties.

I want to say a practical word on physical containment as at present outlined in the NIH guidelines. I say it with deep and special personal misgivings because I come from the biological laboratories at Harvard, where right now is being built a P3 facility on our fourth floor in an old building absolutely infested with little red ants, to such a degree that the food machines have had to be shut off for the last few months because they delivered the bodies of little red ants whenever one tried to draw a cup of tea. So that is where we are putting our P3 facility, and I simply want to point out to you that the NIH guidelines don't ask for the eradication of insects and rodents; they ask for an insect and rodent control program. I want to tell you with great pleasure that Harvard now has an insect and rodent control program. We have both the program and the ants.

I have another question directed to the workshop on the industrial uses and the patents. I have a question to the directors of this whole conference. Who picked the cochairmen for the workshops? I would be interested. But what I really wanted to address to that workshop was a question--or to anyone in the audience who can answer it: What lies behind this sudden and drastic reversal of the position of the pharmaceutical and chemical industry relative to the NIH guidelines? Last November, in a meeting with Ancker-Johnson, one of the cochairmen of this workshop, the industry rejected the guidelines rather arrogantly in the news report I have in hand. The PR men for the industry ended by saying, "It took the scientists two years; we are going to make our own guidelines and it may take us longer."

Now I come back from a little vacation in Peru to find this position completely reversed and the industry joining NIH in asking for legislation. A very curious reversal of position. I deposit that as a serious question.

This prospect of human genetic engineering is a very deep and serious one. The curious thing is that at a hearing before the Cambridge Experimentation Review Board, when it was brought up, my friend David Baltimore said, "Why bring that up? That may be ten years off." Our Peoples Business Commission representative was a little more conservative and gave it twenty years.

I want to say something about my own position with regard to this prospect. Every creature now alive on the earth represents an unbroken line of life that has stretched three billion years, back to the first living organism to appear on the earth. If that line had

ever been broken, how could we be here? I myself will do everything
I can as long as I can to press for a deep principle in law of the
inviolability of the human germ plasm. I want it to have the same
status as our principle of the inviolability of human life, as all of
you know.

There are conditions in which that principle of the inviolability
of human life is overridden, not by terrorists as a rule, but by
government. Instances may come in which one overrides that principle,
but I want the barriers to be high.

HAMBURG: There is a specific question raised which the gentleman here
in front can respond to briefly to answer the question.

RONALD BROOKS, General Electric Company: I was in attendance at the
meeting which was just referred to. There are several members of
industry who are here in the audience who were also there. I would
just like it to go on record that there was a consensus to support the
NIH guidelines that was arrived at at Ancker-Johnson's meeting.

That position was also affirmed in the report on workshop 8 this
morning, and I think that that would be a rather gross exaggeration
as to what happened at that meeting, since I was personally in attendance.

JERRY GLENN: I do contract futures research for the Canadian government,
the U.S. government, corporations, nongovernmental organizations,
and once in a while universities.

My suggestion both to the Academy and to Mr. Ted Howard is that we
take a lesson from the Chinese who have survived as a culture for a
long time, and that is to go from the yin-yang, where you pit one
versus the other and try to resolve it, to the symbol of the universe,
which has three elements in the circle. And what I would suggest
is you set up a physical environment that tracks not only the data,
but the arguments, the polemics, the research, the contracts, etc.,
into three areas--those that want to stop it, those that want to do
it, and those that are not sure. And that you actually have, every
month or so, reports coming out of this place.

Since I do strategic research, for that is really what futures
research really boils down to, I am up to my eyebrows in almost every
controversy facing the world right now, and I agree that I don't want
to have to deal with thirty years of strategic research on this issue.
I would like us to shortcut it by learning from the Chinese.

ROLLIN HOTCHKISS, Rockefeller University: I don't want to enter into the
controversial issues. I would like to speak about two points raised
by the workshop discussions, and one of them is rather general. It
is a matter, I think, of some perspective, and you are going to hear
later from Dr. Tracy Sonneborn. I know that it is unlikely that anyone can say as well as I can how statesmanlike he has been in this
issue, and I am afraid it will not appear unless I say so. As far

back as 1964 he arranged for a symposium on the then just dimly perceived ideas of genetic engineering, and a number of people participated in that, of whom I was one. That appeared in a book in about 1964, and it resulted in some of us there warning about the issues as they were perceived then by the rather primitive transfer of DNA, which has been my other source of remuneration during these years.

The dangers that we foresaw, Dr. Luria and I in particular pointed to these, caused me to go to some extent on the lecture platform, and in 1965 I wrote an article in which I pointed out a number of these things. I am not trying to recommend that article to you so much as to say there were serious, conscientious reactions of a group of people then appearing in books, appearing in lectures, and that was followed, in my case, by going to several schools, science schools, high schools, Catholic high schools, groups of businessmen and trying to see how interested they were or were not in genetic engineering as we perceived it then, and we perceived it with much of the same aura of science fiction, enormous possibilities and unknown risks. My own prescription then was that we should have informed discussion, have serious, responsible discussions among the experts, and that prescription has been so well fulfilled in these last years I have not even felt impelled to get to the microphone before now.

I want to say, then, that I am pleased, delighted with what is going on. The NIH guidelines as an ongoing thing that can be improved from time to time, this type of discussion, and the only thing I regret is that adversary procedures seem to play so large a part in it, when informed, conscientious discussion among people with somewhat greater humility would I think accomplish the purpose better.

My next comment will seem to you perhaps on the other side. It is something that carries a warning, a very different point reflecting discussion that could have appeared last night in the workshop cochaired by Dr. Cohen. Some of you may know, but I think very few of you know, that we have been fusing bacterial protoplasts very recently. This has been done in so few laboratories that it has not yet attracted very wide attention. Now this represents a new kind of engineering. It is not recombinant DNA, but it enters into the question several ways. One way is that it does give quite new techniques for introducing recombinant DNA if you have prepared it in one organism or vector. It also suggests very strongly to me that a good deal of exchange must have occurred in nature, and I would have proposed that had I chosen Dr. Cohen's workshop last night instead of the more difficult one of trying to balance risks and benefits.

The bacterial cell, once divested of its cell wall, is surrounded by a membrane which is very much like the membrane of every other cell, and potentially capable of fusing with every other cell. That means not only that bacteria can be fused with each other within the same species, but different genomes of different evolutionary origin can very likely be postulated now to be put together and juxtaposed; and bacterial protoplasts could be supposed to fuse with such things as plant cell protoplasts to introduce the nitrogen-fixation genes that you are going to hear about shortly I hope, if I can make this short.

Then this will, I think, introduce or go hand in hand with the microbial recombinant DNA work, because it will become a useful tool and the only reason it hasn't been much talked about is because it is relatively little known. It is just beginning.

If I may now make my extrapolation, I think there will be a great deal of work soon on the fusion of such protoplasts. But again, let me say that I think it suggests in nature the opportunity for such fusion may well have occurred many, many times, and we may find that the species as they have evolved themselves have already made many of their decisions about which things will be compatible and which ones will not.

Finally, again to commend the chairman of this symposium, as I don't intend to get up again, I think one of the best signs that an ongoing discussion has been going on, and one of the things that seems heartwarming to me is that, for example, when I was warning against the dangers a few years ago Robert Sinsheimer seemed to be on the other side. Now he has had some second thoughts. He is more worried than I am because he doesn't see these discussions as going quite far enough. But that, I submit to you, is a conscientious and rather humble and serious opinion from people who care very much about how this thing is going.

JOHNSON: I would like to make three comments on Dr. Wald's expressions of opinion, and I use that terminology as opposed to expressions of fact.

Perhaps a minor point, I really don't think that all the terrorist actions are taken by people in uniform of responsible governments. I think the events of Munich during the Olympics, the seizing of airplanes, and this sort of thing really do not involve people in uniform.

Secondly, I would like to correct the impression that the members of the Pharmaceutical Manufacturers Association at any time rejected the NIH guidelines. That is categorically wrong. And in testimony in front of Senator Kennedy and in many public expositions we have consistently supported them.

Finally, I would like to comment on this matter of cloning people. This arises again and again, and this is a serious and reasonable concern of many people in this meeting. It is a concern of mine, and if we ever get to that point, and I object to it very strongly, you will find me in the trenches and on the ramparts with Mr. Rifkin and any other people who want to join us. But I don't think that is really the key issue of much of the basic discussion of this meeting.

AUDIENCE: I would like to comment very briefly on Mr. Howard's statements or presentation. I think that he is right that we are going to have to face a serious issue in terms of the question of human genetic engineering, perhaps not for another twenty or thirty years, but I think it obviously is going to be there.

I think he maybe isn't quite entirely right in saying that this is totally unprecedented. It seems to me that there has been a time in in the past when it was attempted to apply an emerging scientific technology to humans and eventually was rejected, and that is the application of the principles of scientific animal and plant breeding to humans in the form of eugenics. Unfortunately, if you look back at that controversy which probably started just about one hundred years ago, one finds that geneticists generally held themselves aloof from the whole controversy, and it was only resolved when the Nazi Germans overextended themselves in terms of applying eugenics, and I hope that we don't make that same mistake again.

DANIEL E. KOSHLAND, JR., University of California, Berkeley: I have two comments. One, I have heard so many times the comment made that the scientists are making a decision, a small, elite group, and I just would like to comment that the Congress of the United States has, I think at the moment, no scientists. The President, I think has had a couple of courses in engineering, and they are the final arbiters of power. It seems to me quite clear who makes the final decisions in our society.

Coming to conferences of this sort, it seems to me we are all in the intermediate phase of trying to influence science policy. The National Academy, under some pressure and I must say great division, has decided to have meetings of this sort. There is a great deal of disagreement. De Tocqueville said many years ago that the people will believe a simple lie in preference to a complicated truth. I am not that cynical, but I do believe it takes a great deal of effort to understand a complicated truth. I think all of these issues that involve the interface of science and society are really in that category, and I must say I do not agree with the argument that there is sort of a separation in the issue of morals and benefits and risks. This is the only one of many of these forums, and really the whole forums can be discussed in benefits and risks, because that is really what we are talking about. If you really say at a certain point, are we going to proceed with this experiment which might have great benefit to the food supply of the world, and it had absolutely minimal risks, I think all of us would say go ahead. If we talk of advantages to the cosmetic industry from making a new cosmetic in an experiment that involved a really great risk, I think all of us would agree we wouldn't go ahead. So this issue of trying to divide it so simply into moral issues, one class of benefit or one class of risk is, I think, greatly oversimplifying the subject.

HAMBURG: I may say that Dr. Koshland, along with Dr. Philip Handler, played a great role in opening up the Academy to this sort of discussion, not just today but on other controversial topics, the large, complicated issues of science, social change, and public policy. I personally think we owe Dr. Koshland and Dr. Handler a debt of gratitude for doing that. The Academy, of course, is a small part of the whole debate, but to whatever extent the Academy can play a role, Dr. Koshland has, I think, augmented its potentiality.

ALAN KAY, NIH:  My comment is about, again, workshop 8.  I was there
from the beginning to the end.  At no time was there any detailed
discussion of guidelines that would be applicable to industry.  There
was no discussion of what sort of important procedures would be used.
So when the report on the workshop was issued, and at the end of
that report there was a so-called consensus about specific types
of things which could be included in the guidelines, and also some
talk about what type of enforcement could be used, I don't think
that that can be true.  If it was not discussed, there can be no
consensus.

One of the main points of the workshop was the fact, to address Dr.
Wald's point about industry acceptance of the NIH guidelines, that
there are no guidelines at the present moment applicable to the scaling up to industrial use of recombinant DNA research.  Workshop 8
was overwhelmingly represented by people from industry.  There were
very few, as far as I could see, academic scientists, and I think
that again was a pity.  But one of the main thrusts there was that
industry should take the lead in developing the guidelines that
would be applicable to industrial research.

After some questioning they said that this should involve discussion with local people, etc., but I still get the impression that a
main part of industrial effort will be toward lobbying to get guidelines which will be agreeable to industry, and which will, in many
cases, be less strict than the NIH guidelines.

I also got the impression from the President of the Cetus Corporation that he also thinks that some members of industry now feel that
direct lobbying of the federal government would be the best way to
get compliant guidelines for industry.  But he thinks that this would
be a mistake, and that in the resulting uproar the viewpoint of industry
would more or less be done completely away with.  And I think in that
respect he is very right.  He represented, during the whole meeting
last night, the liberal business viewpoint.  Others represented more
conservative business viewpoints.

JOSEPHINE SIMONS, Guest Scientist, NIH:  I would like to request a consensus of all research workers that all strains of organisms used
as recipients for new genetic material such as the $\chi$-1776 and any other
organism that may be used in the future be deposited with the American
Type Culture Collection, as well as all clones produced by introduction of new genetic material, and also such new organisms as may
be produced by the fusion of protoplasts.

FRANCINE SIMRING, Friends of the Earth:  I would like to say concerning
the statement of the gentleman from Eli Lilly, who said that industry
has always been going along with compliance to the guidelines, that
industry has indeed not!  On last June 2, 1976, in their meeting
with Donald Fredrickson, they so stated that they could not comply
with some of the requirements of the NIH guidelines.  And on September
22, the Kennedy Health Subcommittee meeting referred to by the

gentleman from Lilly, Dr. John P. Adams, representing the Pharmaceutical Manufacturers Association, went on the record in his testimony as
stating that manufacturers could not work with a ten-liter limit as
defined in the NIH guidelines, and when pressed for a minimum stated
that 1,000 gallons was the minimum he could work with on an industrial
basis.

You can read in *Business Week* of January 17, 1977, that the Cetus
Corporation is having thoughts about scaling up to 50,000 gallons.
There are no limits at the present time among industry people.

One other point, John Adams repeated his testimony, a 1,000-gallon
minimum, at the October 21 hearing at the Attorney General Lefkowitz's
New York State meeting at that time.  Also I would point out one other
thing that may tie in with industry.  The Department of Transportation
has published in the *Federal Register,* and the deadline has already
passed for comment, on the transportation of recombinant DNA materials.
They are asking that it be included under the Hazardous Materials
Transportation Act.

I think it is an important question for all of us to ask ourselves,
if you are going to spend a million dollars for a top containment
security lab, what are we doing in the link of safety, the chain,
isn't this a weak link?  If you are going to truck or train or plane
these materials the requirement is only that it be in a special kind
of double canister that can fall from a height of thirty feet to a
hard surface.  That is the same regulation for shipping plutonium,
which of course was banned through New York City, as we know.

I wish to open this to scientists who would like to comment on
what kind of safety is offered, if we are going to discuss safety
procedures.

## GENETIC ENGINEERING IN AGRICULTURE WITH EMPHASIS ON BIOLOGICAL NITROGEN FIXATION*

Raymond C. Valentine

*Plant Growth Laboratory, Department of Agronomy and Range Science, University of California, Davis*

### GREEN PLANTS AS SOLAR ENERGY MACHINES

Mankind has become increasingly dependent on energy in all its myriad forms to fuel our homes, factories, farms, and automobiles. Unfortunately, the supply of the most common form of energy that we now use-- petroleum--will undoubtedly become depleted in the years ahead. Solar energy, a renewable form of energy, is emerging as one of the most attractive alternatives. Today, we recognize the green plant as a marvelous machine capable of converting sunlight into useful products --food, fiber, and oils. Put simply, if we improve the productivity of plants, we will enhance their utility as solar energy machines, leading to the day when we will be self-sufficient for energy.

*Research on the energy cost of $N_2$ fixation was supported by the National Science Foundation, RANN (Research Applied to National Needs, APR 75-09577).

## GENETIC ENGINEERING, AN ANCIENT ART

Genetic engineering, the application of genetics, is an ancient art in agriculture whose earliest practitioners simply used a keen eye to choose hardy races of plants and animals for their domestication. Although there is still some uncertainty about the precise historical origins of many of our crops and animals, there is no doubt that the art of genetic engineering was practiced even from those early times (helped by the bees). The modern geneticist continues to mold and groom plant and animal life to provide the needs of a hungry world.

The 1970s has brought a new tool for genetic engineering in agriculture--recombinant DNA. There are many who feel that this new technology offers a quantum jump for improving our crops. Undoubtedly, research on recombinant DNA will have great benefit to society through application to agriculture, perhaps even more so in the long run than in medicine. Many areas come to mind in which genetic engineering with recombinant DNA offers great promise:

- Enhancement of biological nitrogen fixation
- Photosynthesis and increased efficiency of carbon dioxide fixation
- Biological pest control
- Fuel production (hydrogen, methane, etc.) through bioconversion
- Plant breeding

## POTENTIAL OF GENETIC ENGINEERING OF NITROGEN FIXATION

It is widely recognized that available nitrogen limits the productivity of much of the world's agricultural land. This has led to the extensive use of chemical nitrogen fertilizer, a practice that has been so successful that it has become synonymous with the much-used term "green revolution."

Today, we are faced with one of the greatest challenges of our time. How can we continue to maintain crop yields with nitrogen fertilizer when its cost of production is increasing due to depletion of non-renewable fossil fuels? This dilemma has led to worldwide effort to enhance biological production of nitrogen fertilizer, a process which uses solar energy, a renewable energy form, rather than an ever-shrinking fossil fuel supply.

Only bacteria and blue-green algae have evolved the capacity to fix nitrogen (atmospheric nitrogen gas [$N_2$] $\rightarrow$ ammonia). Certain green plants such as the legumes (e.g., soybeans, alfalfa, clover, peanuts) depend on symbiotic associations with nitrogen-fixing bacteria for supplying their N-fertilizer.

The well-nodulated soybean shown in Figure 20 represents an important example of symbiotic nitrogen fixation. The nodules that are visible, as ball-like structures on the roots, behave as little factories for manufacturing the plant's own supply of N-fertilizer using energy from the sun. In contrast, our important cereal grains such as corn and

FIGURE 20   A well-
nodulated soybean.

wheat lack nodules and must be supplied with massive doses of commercial
N-fertilizer.

The biological production of N-fertilizer is governed by a set of
genes which we named the Nif genes.[1]  It is now possible to manipulate
the nitrogen-fixation genes using the techniques of genetic engineering.
The two major goals of this research are:

- Enhancing the efficiency of nitrogen fixation by natural systems
(e.g., soybeans, alfalfa, and peanuts)
- Construction of new nitrogen-fixing plants (e.g., corn, wheat)

Although these are exciting prospects with great potential benefit to
agriculture, it is important to carefully weigh all possible risks.
Would genetically engineered, nitrogen-fixing organisms pose a threat
to the environment?  There is now considerable information in the scientific literature leading to the conclusion that the fixation of nitrogen
requires a tremendous amount of energy, thereby placing severe energy
constraints on organisms with this trait.  According to this hypothesis,
most natural environments (soil, water, etc.) simply lack sufficient
energy to sustain a runaway nitrogen-fixing organism--a sort of natural

barrier which blocks hazardous overproduction of fixed nitrogen. To test this idea, we have constructed and studied special mutant strains of a soil bacterium called *Klebsiella* which excretes large quantities of fixed nitrogen into its environment in the laboratory.[2,3]

The strains of *Klebsiella* used in these experiments are called "nitrogenase derepressed" because they have been mutated by conventional techniques to continue to synthesize the nitrogen-fixing enzyme, nitrogenase, in the presence of an excess of ammonium ion, the product of nitrogen fixation.[4] In contrast, the parental (wild-type) organisms are very sensitive to the accumulation of ammonium ion, nitrogenase production being completely shut down in the presence of small quantities of ammonium in their environment. Thus, fixed nitrogen does not accumulate with the wild-type strain. In contrast, when Nif-derepressed mutants are grown inside a dialysis bag (Figure 21) that is suspended in a reservoir of glucose as energy source, large quantities of fixed nitrogen are excreted.

The Nif-derepressed *Klebsiella* bacteria multiply rapidly at first, using up their growth-limiting supply of amino acids, at which time growth ceases. Essentially all of the fixed nitrogen is exported during the stage when the cells have stopped actively dividing (stationary). During the active period of nitrogen fixation by these cultures, which may last almost one week with certain strains, the consumption of energy (in this case, the simple sugar glucose) was determined as a function of ammonium produced[5] (Figure 22). The ratio of glucose consumed as energy source per ammonium produced has been found to fluctuate widely depending on environmental conditions, with the most efficient stage of nitrogen fixation requiring eight to ten glucose per nitrogen gas reduced ($N_2$). In further studies of the energy cost of $N_2$ fixation *in vivo*, Andersen and Shanmugam in my laboratory have calculated that about twenty-one to twenty-five ATPs (the most common energy currency of the cell) are needed for each $N_2$ converted to ammonia by *Klebsiella* (Table 4). This molar ratio of ATP/$N_2$ is in general agreement with an ATP/$N_2$ molar ratio of

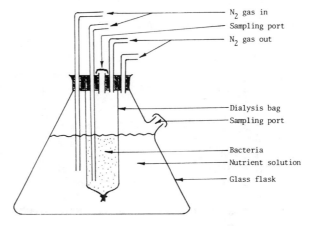

FIGURE 21 Artificial nodule for evaluating the energy cost of $N_2$ fixation. Nitrogen-fixation (Nif)-derepressed *Klebsiella* in an experimental apparatus continue fixing nitrogen after their own protein needs are satisfied and release excess ammonia to the environment. A glucose solution provides the energy needed. See text for further details.

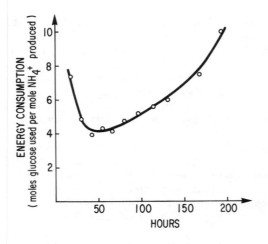

FIGURE 22 Energy cost of $N_2$ fixation *in vivo* in terms of glucose consumed per $NH_4^+$ fixed from $N_2$. See Figure 21 and text for details.

TABLE 4 Energy Requirement for $N_2$ Fixation *in vivo*

| Organism | Moles of ATP Required per Mole $N_2$ Fixed | Method of Calculation | Reference |
|---|---|---|---|
| *Clostridium pasteuranium* | 20 | Cell yield | 6 |
| *Azotobacter chroococcum* | 4-5 | Cell yield | 8 |
| *Klebsiella pneumoniae* | 29 | Cell yield | 7 |
| *Klebsiella pneumoniae* | 21-25 | $NH_4^+$ production | 5 |

twenty reported by Daesch and Mortenson[6] for *Clostridium* and a value of twenty-nine by Hill[7] for *Klebsiella* using a different procedure from ours (see Table 4). We have no explanation for the value of four to five ATP/ $N_2$ reported by Dalton and Postgate[8] for *Azotobacter*, but feel that their value is unrealistically low and should be reexamined.

We concluded from these experiments that biological nitrogen fixation by the soil microorganisms *Klebsiella* and *Clostridium* is an extremely energy-intensive process requiring large quantities of metabolic energy.

Thus, availability of energy provides a major restraint to contamination of the environment by genetically engineered soil organisms.

In the case of symbiotic $N_2$ fixation in legumes, such as soybeans, energy supply has also been identified as a cardinal rate-limiting step for the process.[9] Once again, harmful accumulation of fixed nitrogen is prohibited by lack of available energy. Experiments are in process to determine if similar energy constraints occur with all nitrogen-fixing organisms.

## THE FERTILIZER OZONE PROBLEM

Both chemical and biologically fixed nitrogen may contribute to the synthesis of oxides of nitrogen, which in turn may be destructive of our "ozone shield" protecting us from harmful ultraviolet rays. Thus, enhancing biological nitrogen fixation may result in some increases in production of oxides of nitrogen. However, there are at least two points which favor biological synthesis of N-fertilizer such as occurs in leguminous plants over the chemical way: (1) nitrogen produced during symbiotic $N_2$ fixation is made available slowly to the plant, permitting all of the fixed nitrogen to be consumed within the plant; and (2) nitrogen tied up in the plant is simply not as readily available for conversion to oxides of nitrogen as is the case with added commercial fertilizer. Thus, biologically fixed nitrogen may be a relatively clean process with respect to production of oxides of nitrogen.

## RESEARCH PRIORITIES

The cardinal rule for evaluating research priorities in the field of biological nitrogen fixation concerns the use of radiant energy to fuel the process. Many of our most successful crops such as soybean, alfalfa, clover, peanuts, and peas already have evolved this crucial capacity. Research on these crops should be intensified. It may be possible to harness new naturally occurring systems.

For instance, an interesting and potentially valuable example of using solar energy to produce N-fertilizer is illustrated for the tiny water fern *Azolla* in Figure 23. This fern is used by farmers in the Far East for producing nitrogen fertilizer for their rice paddies. *Azolla* grows luxuriantly in rice paddies, forming a nitrogen-rich mat on the surface of the water among the rice plants. *Azolla* is able to manufacture its own nitrogen fertilizer because of a symbiotic relationship with blue-green algae. The blue-green algae live in a leaf pouch of *Azolla* and convert atmospheric nitrogen gas into a usable form of nitrogen for the plant. Nitrogen compounds from *Azolla* become available to the rice plant when *Azolla* decays in the soil. There is some potential that *Azolla* might be used to produce N fertilizer as illustrated by Figure 23. In this scheme, varieties of *Azolla* that produce nitrogen fertilizer as ammonium ($NH_4^+$) are cultivated in a separate pond (or in the case of rice, within

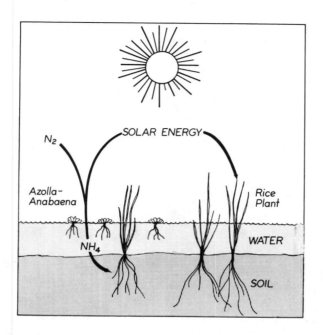

FIGURE 23 Theoretical scheme for biological production of N-fertilizer by *Azolla* in rice culture.

the same paddy), which is connected to the irrigation system, allowing nitrogen to be flushed from the *Azolla* pond into the field or orchard through the irrigation water. In the future, it may be possible to genetically engineer the *Azolla*/blue-green algae system to export increased levels of N-fertilizer.

One of the great future challenges for biologists is to genetically engineer new nitrogen-fixing plants. This is an extremely difficult task because of the many requirements that first must be met such as protecting the system against oxygen damage, providing sufficient energy, etc. It is a controversial point whether genetically engineered crops such as corn and wheat that produce their own N-fertilizer would be at a disadvantage because of the large amount of energy they must invest in this new process. In trying to evaluate the impact of nitrogen fixation on new crops, it should be kept in mind that most crop plants use the nitrate form of N-fertilizer; this process also requires a lot of energy. Thus, the additional burden that a new nitrogen-fixing crop must bear may not be as large as many scientists believe.

Unfortunately, recent reports concerning the potential of biological nitrogen fixation in cereal grains appear to have been grossly exaggerated, leaving no cleancut case of a beneficial, naturally occurring symbiosis with cereals.


CONCLUSION

Genetic engineering of the nitrogen-fixation genes holds great promise for enhancing crop productivity; however, it is essential that both the benefits and risks of such technology be weighed and analyzed carefully in order not to disturb other essential natural cycles (e.g., ozone shield).

It is now clear that massive quantities of metabolic energy are required for biological nitrogen fixation. Consequently, limitation of available energy (as photosynthate in plants and organic compounds for heterotrophic bacteria) appears to strongly prohibit nitrogen-fixation genes from causing damage to the environment. Also, biological nitrogen fixation may turn out to be a relatively clean process with respect to synthesis of oxides of nitrogen. Therefore, it seems reasonable to proceed with genetic research aimed at enhancing biological $N_2$ fixation.

However, the risks as well as the benefits of other aspects of genetic engineering in agriculture must be weighed carefully before proceeding. The agricultural community as well as society in general must be party to these decisions.

## REFERENCES

1. Streicher, S., E. Gurney, and R. C. Valentine. 1971. Proc. Natl. Acad. Sci. U.S.A. 68:1174.
2. Shanmugam, K. T., and R. C. Valentine. 1975a. Science 187:919.
3. Shanmugam, K. T., and R. C. Valentine. 1975b. Proc. Natl. Acad. Sci. U.S.A. 72:136.
4. Shanmugam, K. T., I. Chan, and C. Morandi. 1975. Biochim. Biophys. Acta 408:101.
5. Andersen, K., and K. T. Shanmugam. 1977. *In* A. Hollaender (ed.), *Brookhaven Symposium on $N_2$ Fixation*. In press.
6. Daesch, G., and K. E. Mortenson. 1968. J. Bacteriol. 96:346.
7. Hill, S. 1976. J. Gen. Microbiol. 95:297.
8. Dalton, H., and J. R. Postgate. 1969. J. Gen. Microbiol. 56:307.
9. Hardy, R. W. F., and U. D. Havelka. 1975. Science 188:633.
10. Delwiche, C. C., and B. A. Bryan. 1976. Annu. Rev. Microbiol. 30:241.
11. News and Comment Science. 1977. 195:658.

## RECOMBINANT DNA: IT'S NOT WHAT WE NEED

Ethan R. Signer

*Professor of Biology, Massachusetts Institute of Technology*

People are going hungry.

Let's take this country, ignore the junk food, the additives, the synthetics, even ignore all the trimmings and leftovers we just throw away.

At least 25 million Americans are below the poverty line, $5,500 a year for a family of four,[1] hardly enough for a decent diet. Thirty-five million Americans or more are eligible for food stamps.[2] Some people are so poor they eat dog food. We're the richest nation in history, and we can't feed all our people.

Internationally, the World Bank estimates[3] that 75 percent of the population of the so-called "underdeveloped" countries are getting too few calories. That's more than a billion people, a quarter of the earth's population. Yet we could feed them all if we were to increase the world annual grain production by only 4 percent.[3] Can we really not turn out another 4 percent?

No, there's plenty of food, but not everybody gets it. Some people don't have the money; some nations don't have the power.

Recombinant DNA research might increase nitrogen fixation years from now, though other methods are just as likely to. But that won't put any more food into people's mouths. Look at the last technological miracle, the Green Revolution. It's been a failure. It hasn't relieved hunger in Asia, it's only made the rich richer and the poor even poorer.

The problem is political, not technological, and it's going to be with us until there's a political solution. People are hungry, but it's not for lack of recombinant DNA research. That's not what we really need.

Medicine is the other area that's supposed to benefit, and the same is true. Perhaps we are expanding the frontiers of medically related research. But what people, particularly working people, are actually getting now is much less than what they could already.

We haven't enough doctors. Hospitals are closing down. Treatment costs more and more. Occupational health hazards are ignored. Drug companies milk us for profits. Health care gets less and less humane and dignified, and we all know there's one standard for the rich and an even lower one for the poor.

Perhaps we shouldn't be surprised. How concerned can we be for the health of our people when we don't do much about preventing disease in the first place? Take cancer:

"It is now clear...that most if not all cancers have environmental causes and can in principle be prevented."[4]

Simple common sense says that we solve problems by first trying what we know how to do, especially when it might work.

Do we want to fight cancer? Then it's not recombinant DNA research that we need. Or, if we want to do the research, let's not try and justify it as a magic cancer remedy when there is one already. Is it insulin we're talking about? Hog insulin works fine, let's grow more pigs, and let's cut drug prices before we start thinking about miracle cures.

These are the bottlenecks in making our people healthier, not a lack of recombinant DNA research. And any benefits it does yield won't reach the public a bit more effectively than what we have now. What's more, we might well get them from other research.

In health as in food, research should address what we really need, and this isn't it.

On the other hand, there are the risks.  Recombinant DNA research will create new biological forms.  They'll be able to duplicate themselves, and nobody can be sure how some of them will behave.  The doomsday scenarios are pretty frightening.  Unlikely, perhaps, but one in a zillion is still "one," and one disaster that size is a pretty big "one."

But there's much worse:  What about the scenarios we *haven't* thought of?  Five years ago we hadn't even thought of recombinant DNA technology!  How do we weigh benefits and risks when we can't even guess at some of the risks?

Paul Berg (Stanford University), a leading advocate of this research, says, "...we shall be doing things that would have been thought completely improbable a few years ago."[5]  James Watson (Director, Cold Spring Harbor Laboratory), another leading advocate, is explicit:  "We can't even measure the...risks!"[6]

So not only don't we really need it, but there are risks we can't even judge.

Now, it's said that we can't limit free inquiry.  But we already do.  We don't permit vivisection, or experimenting on people who don't consent to it.  So recombinant DNA wouldn't be a special case, just another item on the list.

Some say we have to do this research because it's new and exciting and challenging, because it's progress.  That's really tempting.  The human spirit is at its finest up against the unknown.  But not when we need this so little, not when we risk such enormous damage.  We're already going to the moon and Mars and Venus, that's challenge enough.

Sometimes we hear that this is simply part of the basic research society has to have.  Maxine Singer (National Institutes of Health) and Paul Berg, both leading advocates, say that "the only certain benefit is increased knowledge of basic biologic processes."[7]  Notice, though, "the *only* certain benefit."

We do need to search for pure, basic, unpredictable knowledge.  But we can't have it both ways.  If it's unpredictable knowledge we want, then we can search in any direction.  There's a whole biosphere to study; recombinant DNA is hardly the only thing left.  If we say this direction is more important than any other, then it's no longer a case of pure knowledge and we're back to benefits--only speculative, not certain--that aren't what we really need.

But even that's really beside the point.  Nobody's trying to limit the search for knowledge.  Recombinant DNA is only one technique, and that's all.  If it is the structure of the gene we absolutely must know about-- well, we molecular biologists are brilliant.  Surely we can figure out how to study it without using recombinant DNA.  Now there's a challenge!

And why the big rush?  Maybe a University of Michigan professor had the answer:  "...other universities will proceed with...research on this subject.  Should Michigan choose not to, we will lose our position...."[8]  Not much of a justification for taking risks.

Now, there are scientists pro and con, but you don't hear a public hue and cry demanding recombinant DNA research.  In fact, most people

234

are frightened of it. Figure 24 shows how the *Boston Globe* cartoonist pictured Cambridge's decision to allow this research.

We need the public. This research will eventually tinker with the gene pool of humanity. So the public, like the subject of any experiment, must give its informed consent--but willingly, not by coercion. Let's not have decision making turn into self-serving professional deception, as in the American Medical Association, whose number everybody must certainly have by now.

Of course, that's playing with fire. The public might not want this research. Maybe they suspect that technology serves only its masters, and not the public at large. Maybe they'll take the common sense approach to any risky course of action, and put the burden of proof on the advocates, rather than on the opponents. That's not irrational fear of the unknown-- it's rational, sensible fear of the unknown. People are not crazy to be scared.

At Asilomar, Sydney Brenner (Medical Research Council Laboratory, Cambridge, England), a leading advocate, cautioned that we scientists have to "...reject the attitude [of let's]...pretend there's a biohazard and hope we can arrive at a compromise that won't affect my small area, and I can get my tenure and grants and be appointed to the National Academy and all the other things that scientists seem to be interested in."[9] And Roger Dworkin (Indiana University) warned that "Any appearance of self-serving will sacrifice the reservoir of respect that scientists have and will bring disaster on them."[10] Why, then, are the important decisions about regulation being made by those very scientists who are going

'Crack out the liquid nitrogen, dumplings . . . we're on our way!'

FIGURE 24   Reprinted from the *Boston Globe*, February 9, 1977, by permission of McNaught Syndicate.

to do the research?  It's true we scientists know all about the technical details, but this is not an issue of technical expertise.  Scientists don't necessarily know best.  People are smart enough and informed enough to decide for themselves whether to let scientists plunge on with a risky technique that's not what we need, whether to let scientists tamper artificially with genes that could soon be our own.

We shouldn't ignore a few of the bigger, more certain risks.  One is military.  We know their record.  They'll use anything as a weapon that they can get their trigger fingers on.  Public opinion or official policy won't stand in the way of classified research.

Another is industry. Seven of our biggest drug companies are already doing or planning recombinant DNA research.  And we know the drug industry --outrageous profits, competitive pressures, secrecy for patent rights, contempt for the public.  Corporations have even been formed by scientists specifically to exploit recombinant DNA.  One of them, the Cetus Corporation, acknowledges that, "It is...still difficult to find any really important medical or industrial capability for which it matters at all that we know the genetic code...."[11]  Yet a few pages later--"We are proposing to create an entire new industry to [focus] on those specific problems that appear most amenable to solution...and promise the best cost-benefit ratio."[12]  That's awfully close to saying, "It looks like nobody's going to need this, so we'd better find a use for it, and besides, we'll make a lot of money."

Finally, and by no means least, recombinant DNA research is going to bring us one more step closer to genetic engineering of people.  That's when they figure out how to have us produce children with ideal characteristics.  Last time around, the ideal children had blond hair, blue eyes, and Aryan genes.  I can hardly wait.

Let me sum up:  In food, as in health, we're not doing what we already could.  Recombinant DNA won't change this.  We ought to balance the risks, not against speculative benefits, but against need.

This is not what we need, it's a luxury.  Furthermore it's risky, maybe even incredibly risky, to us, to the biosphere, to the future.  We're going to tinker with the human gene pool.  We don't need it, it's risky, it's certain to be abused; common sense says we ought to stop doing it.  So let's stop--now.

NOTES AND REFERENCES

1. United States Census Bureau.  1976.  Annual Report.
2. Generally accepted figure among U.S. government agencies.
3. *Malnutrition and Poverty*, Johns Hopkins University Press, Baltimore.
   Quoted in Science 194:1142, 1977.
4. Doll, R.  1977.  Nature 265:589.
5. Berg, P.  1976.  ASM News 42:273.
6. Watson, J.  Quoted in Rolling Stone, June 19, 1975, p. 42.
7. Singer, M., and P. Berg.  1976.  Science 193:188.

236

8.  Unidentified University of Michigan professor.  Quoted in The Atlantic, February 1977, p. 59.
9.  Brenner, S.  Quoted in Rolling Stone, June 19, 1975, p. 40.
10. Dworkin, R.  Quoted in Science 187:935, 1975.
11. Cetus Corporation, Special Report, p. 2.
12. *Ibid.*, p. 8.

## POTENTIAL USES

David Baltimore

*American Cancer Society Professor of Microbiology, Center for Cancer
Research, Massachusetts Institute of Technology*

We have heard in this Forum frequent reference to genetic engineering.
We also have heard that genetic engineering is equal to recombinant DNA
research.  That is not true.

They are very separate concepts.  They must be handled intellectually
as different problems.  I would define genetic engineering as the insertion of new genes or replacement of the defective genes in cells of higher
organisms, especially man.  Genetic engineering can be done in two ways.
It can be done so that the genes of cells in one individual in a population are changed, but his or her offspring do not have an altered genetic
constitution.  That would be a gene therapy for an individual's problems
but would not change the gene pool of the species.

Genetic engineering could, in principle, also be done so that the
altered genome is transferred to offspring, and then to all future generations.  That, I must say, is a form of genetic engineering for which there
is very little prospect in anything like the near future, although the
former kind of genetic engineering I think is much closer.

It is important to remember that genetic engineering is not now a
reality, but is a possibility for the future.  Five years ago it was
often argued that genetic engineering is so far off that we should not
worry about it, but we should put our attention to more immediate

237

concerns. Recombinant DNA techniques have certainly altered that perspective, because many of the stumbling blocks in the way of doing genetic engineering have been removed with the advent of these new methods.

Genes from humans can now be purified and can be inserted into a variety of vectors that could carry them back into the chromosomes of cells. So it is very appropriate that the debates about recombinant DNA also consider genetic engineering.

There are still many people who do not believe that genetic engineering is feasible, so let me offer to you a possible scheme to indicate how close we could be to attempts at genetic engineering. There exist numerous inherited diseases that result in abnormal hemoglobin formation. We have heard discussion of sickle-cell disease and of thalassemia. It is hard to imagine changing the probability of transmission of such genes to people, but we could imagine genetic engineering being used by physicians to treat afflicted individuals. A possible protocol for gene therapy could be something like the following.

One could remove a sample of bone marrow from an afflicted individual, the bone marrow being the site of red-blood-cell formation and the initial site of hemoglobin formation. We could add to the cells the genes of normal globin synthesis, or even regulatory genes if we understand them. These could be attached to some appropriate vector that would help to insert the genes permanently into the cells. The bone marrow would now be easily taken back into the individual because there would be no immunologic barrier, so the bone marrow could be reinserted into the patient, possibly after irradiation to reduce the function of the mass of abnormal cells. Such an operation could be fast, relatively painless, and not very expensive.

I have little doubt that within five to ten years just such an experiment will be attempted, and that, if it is successful, gene therapy could be added to the arsenal of hematologists. I should point out that this scenario assumes that the newly added gene will function normally and under appropriate regulation in the cell which receives it. There is certainly no guarantee of this, but it seems likely that animal experimentation could teach us how to provide the genes in an appropriate manner.

It also seems likely that the cell that received a new gene would have dual function, of both the added gene and the resident defective gene, but that would probably be an acceptable situation from a therapeutic point of view.

I should also point out that this will generate something of a new industry if it comes about, because there will have to be people who know how to make the genes, to link them, and to provide them in therapeutically useful form.

When such therapy becomes possible, there is little doubt that afflicted individuals will seek it. And not to make it available, if it is a feasible scheme, seems inhumane to me. In general, genetic diseases are one of our most serious medical problems, and if gene therapy could be used, many lives could be enriched by better health.

There are certainly other ways to handle the problems that arise from genetic defects, diet being one of them for many types of defects, but medicine is best served by a multipronged attack on all possible levels, and I think the gene therapy will be one.

On a world scale there are certainly much more severe problems than inherited disorders, but to those who are afflicted that fact has very little force. Furthermore, we cannot prevent genetic disorders except by hugely expensive screening programs, and such programs often cover only the further pregnancies of a mother who has already produced a child with some rare genetic defect. So there will be, for the foreseeable future, many genetic problems requiring forms of treatment.

If genetic engineering could be so beneficial, why is it so frightening? Why is it often portrayed as a specter of the future, as we have heard today, rather than as a hope of the future? I think first of all the major fear of genetic engineering is that the "gene pool" of human beings will be tampered with. That is only true if the genes are transmitted from one individual to another, which requires a form of insertion of genes into the "germ line," the germ cells or the sex cells of individuals. That is a much more difficult problem than the type of scenario I have described.

I think there is also the fear that genetic engineering, when it comes about, will provide a methodology that not only could ameliorate the effects of defective genes, but could be also used with what one might call more controversial genes. For instance, if we had a gene that encoded the synthesis of a natural tranquilizing substance, for instance one ordinarily made in the brain, physicians might try to treat certain mental illnesses by inserting such a gene into critical cells. But such a technique could also be used on prisoners, or on anyone considered too far deviant from the norms of a given society.

It is such powers, as hypothetical as they are, that frighten us. Rather than designing increasingly elaborate scenarios for you, I would like to simply pose the central question raised by this speculative look into the future. Should we forego the benefits because of the possible misuses? In considering genetic engineering this question is even more difficult than in our earlier discussions in this meeting, because the concepts of appropriate use and of misuse are extremely difficult to define.

Given the potential for use and misuse, what should we do? We could say, as some have, that genetic engineering is too powerful a tool for any government to be able to control, and so it should not be developed. It could then be argued as many have in this meeting, that because of its implications for genetic engineering recombinant DNA research should be stopped. Remember that recombinant DNA research is only an element of genetic engineering, and has its own applications quite separate from genetic engineering. If we did that it would certainly put genetic engineering into a much more distant future.

Let me take an equally extreme position from an opposite perspective. We could argue that we should simply plunge into genetic engineering as we have plunged into so many other technologies, put no restraints on it, and hope that future generations will be able to cope with the enormous problems inherent in such a powerful technology.

My own answer to these thorny problems of benefits and risks is the same now as when our first ad hoc committee considered how to control recombinant DNA technology. Action of almost any sort implies risk. We should try to identify risks and minimize those we can see. We should be prudent and vigilant in the development of new techniques. But we should not let outselves be frozen by fear of the unknown.

I also believe that controlling future applications of a technology by halting its development is shortsighted. If genetic engineering has a promise of alleviating otherwise untreatable diseases, we have an obligation to make it available. But we also have an obligation to see that the technique is used wisely. Now is certainly none too soon, and it is by no means the start. As Rollin Hotchkiss pointed out, the discussion that began at least thirteen years ago was the start. We should be thinking about these problems. We should be seriously considering the implications of techniques as they develop, and devise methods for controlling their use.

Although scientists, because of their technical knowledge, have a special responsibility to consider the implications in genetic engineering and the necessary controls of it, the responsibility for controlling the fruits of science falls on the total society. The concept of humane use of scientific technology requires definitions of what is humane. Such definitions can only come from the joint efforts of scientists and the lay public.

We are then left with a last, great challenge, defining the public. I personally have been appalled at this meeting and elsewhere by the limited knowledge and poor logic of those representing themselves as the public. This is in direct contrast to a specific exercise in which what I would consider truly public representation considered these problems. That was the activities of the Cambridge City Review Board, where a series of citizens, chosen at random, considered the problems and came up with a measured, intelligent, and deeply understanding response. I hope that, as questions of how to proceed in the future are raised, there will be groups of that sort to join with scientists in thinking out the implications of progress.

## PRESENT AND FUTURE ABUSES

Jonathan Beckwith

*Professor of Microbiology and Molecular Genetics, Harvard Medical School*

I have been doing research in bacterial genetics for the last twelve years at Harvard Medical School, and I am a member of Science for the People.

Over the last couple of years, we have been discussing in our laboratory how the recombinant DNA technique could make certain of our experiments much easier to do. However, as a result of these discussions we decided not to use this technique at all. This is not because the particular experiments we were talking about could be thought of as health hazards in any way. Rather, my reasons were that I do not wish to contribute to the development of a technology which I believe will have profound and harmful effects on this society. I want to explain why some of us have arrived at this decision.

In 1969, a group of us in the laboratory developed a method for purifying a bacterial gene. We took that opportunity to issue a public warning that we saw developments in molecular genetics were leading to the possibility of human genetic engineering.[1] While we saw genetics progressing in this direction, we had no idea how quickly scientists would proceed to overcome some of the major obstacles to manipulating human genes. The reports on the use of recombinant DNA technology, beginning in 1973, represented a major leap forward. The result is that geneticists are now in a position to purify human genes. And proposals have already been put forward for the setting up of "mammalian DNA banks."[2] Further, techniques are being developed that will allow reintroduction of those genes into mammalian cells. These steps appear perfectly feasible.

There are still some barriers left to introducing genes into human cells, organs, or embryos at the proper time or in the proper way. But these goals are not at all inconceivable and they may be achieved very rapidly.[3] Whatever the current state of knowledge, to claim that the possibilities of genetic engineering of humans with this technique is far off is to totally ignore the history of this field.

In 1969, most scientists pointed to the impossibility of purifying human genes and claimed that such developments were at least decades off. In fact, they were four years off. Let's not be fooled again. Just as suddenly as recombinant DNA appeared on the scene, breakthroughs in "genetic surgery" may appear.

And when the day arrives in the near future when geneticists have constructed a "safe" vector for carrying mammalian genes into human cells, others will begin to use it for human genetic engineering purposes. There has already been at least one reported case in which there were direct attempts to cure a genetic disease in human beings with virus-carried genes,[4] and in human cells.

But, why be concerned about human genetic engineering? There are certainly many individuals and groups that have ethical or religious objections to any intervention of this kind in human beings. Possibly after widespread discussion within a society those objections might predominate. I, personally, do not necessarily view all human genetic intervention as inherently to be opposed. But, I would rather point today to some concrete dangers of the development of recombinant DNA research by examining the scientific, social, and political context in which it is proceeding. For that reason, much of what follows will speak to those issues rather than directly to recombinant DNA.

## SCIENTIFIC DEVELOPMENTS

In the last ten or fifteen years, there have been advances in a number of areas of genetics which bring us to a situation today, in which genetic engineering is already under way.  These include a variety of types of genetic screening programs in which it is possible to identify genetic differences between people by examining cells of individuals.[5]  The approaches are:

   1.  Amniocentesis--where the cells of a fetus obtained from a pregnant woman can be examined for genetic variations.  In a small number of cases, these variations are known to cause serious health problems, and suffering may be eliminated by giving the parents the option of aborting such fetuses.
   2.  Postnatal screening--when infants are screened after birth for genetic differences.  Again, in a small number of cases, those variations may cause disease and treatment may be provided.
   3.  Adult screening--where prospective parents can be advised of the likelihood of their bearing children who might carry particular genetic variations.

While each of these programs has proved beneficial to some individuals, they have also encountered problems, been controversial, and, in some cases, caused suffering to those screened.  In addition, all of these programs raise the basic question of who is deciding who is defective, or even, who shall live?[6]
   There are other developments which have received much attention in the press, e.g., the possibility of cloning genetically identical individuals and the attempts to grow fertilized eggs in the test tube and then implant them in a woman's uterus.
   At the same time that these developments in genetic technology were taking place, there was also a growth in studies in human behavioral genetics.  In the last ten years, there has been a resurgence of supposedly "scientific" research that claims to explain many of our social problems as being due to genetic differences between people.[7]  For instance, there are the attempts to say that the inequality which exists in this country or the lower achievement of various groups, particularly blacks, is due to inferior genes.[8]  Or the proposals that criminality might be explained by genetic differences between the criminal and the noncriminal --the case of the XYY male.[9]  (By the way, one of the reasons that I suggest that genetic engineering is already under way, is that XYY fetuses have been aborted after detection by amniocentesis.[10])  In both these cases, the scientific evidence has been shown to be nonexistent and, in some cases, fraudulent.  In addition, there are the more recent attempts in the field of sociobiology to claim biological and genetic evidence to justify the lower status position of women in this society.[11]  It is a disgrace that this government continues to support such shoddy, groundless, and ultimately harmful research.

## SOCIAL-POLITICAL CONTEXT

These genetic theories and the problems with genetic screening programs did not arise in a social and political vacuum. They have followed a period of intense social agitation and social disruption in the United States. After blacks, other minority groups, the poor, and women demanded a greater share of the wealth and power in this society, the response arose that such equality is genetically impossible. The ghetto uprisings and other violent confrontations that occurred during this period are explained as being due to people whose genes are "off." The demands of the women's movement are met with the answer that women are genetically programed for the roles they now occupy.

Another more recent example of this genetic approach to social problems lies in the field of industrial susceptibility screening.[12] Arguments have been appearing in the scientific literature and elsewhere that occupational diseases, caused by pollutants in the workplace, can be ascribed not to the pollutants themselves, but to the fact that some individuals are genetically more susceptible to the pollutants than other individuals. So the argument goes, the solution is not getting rid of the pollutants, but rather, for example, simply not hiring those individuals who are thought to carry the genetic susceptibility. Now, clearly, whenever it is possible to warn someone of dangers he or she may face, that information is important. However, what is blatantly ignored by those promoting this area of research is that, in almost every case, nearly everyone in the workplace is at some degree of increased risk because of exposure, for instance, to asbestos fibers. Yet, already, there are headlines in the newspapers such as the following: "Next Job Application May Include Your Genotype."[13] A Dow Chemical plant in Texas has instituted a large-scale genetic screening program of its workers.[14] Rather than cleaning up the lead oxide in General Motors plants, women of child-bearing age are required to be sterilized if they wish employment.[12,15] It is a genetic cop-out to allow industries to blame the disease on the genetically different individual rather than on their massive pollution of the workplace and the atmosphere. This is the epitome of "blaming the victim."[16]

The end result of these genetic excuses for society's problems is to allow those in power in the society to argue that social, economic, and environmental changes are not needed--that a simpler solution is to keep an eye on people's genes. And thus the priorities are determined. For example, major funding goes to genetics research and into viral causes of cancer, and a pittance to occupational health and safety. This distorted perspective is reinforced by the emphasis and the publicity that recombinant DNA research has achieved with its claim for solving problems whose solution are mainly not in the realm of genetics. Typical of the claims made by those promoting this area is a statement by my fellow panelist David Baltimore: "How much do we need recombinant DNA? Fine, we can do without it. We have lived with famine, virus and cancer, and we can continue to."[17]

That is not a neutral or apolitical statement. The sources of famine and disease lie much more in social and economic arrangements than in lack of technological progress. Aside from the incredible claims for the benefits of recombinant DNA, this statement essentially opts for the status quo. Social problems, such as famine and disease, are taken out of the arena of political action and sanitized behind the white coat of the scientist and the doctor. Of course, we might have both social and medical approaches to such problems going on at the same time. But given the current struggle over solutions to these problems, such statements can only provide weapons to those who would like to maintain present power relationships and profits. What is opted for are the technological fixes, in this case, the genetic fix.

RECOMBINANT DNA--THE GENETIC FIX

Let me give you some examples of how we may move from the present technological fix to the genetic fix, once recombinant DNA techniques have provided the tools. In the United States over the last few years, approximately 1 million school children per year have been given drugs, usually amphetamines, by the school systems, in order to curb what is deemed disruptive behavior in the classroom.[18] It is claimed that these children are all suffering from a medical syndrome, minimal brain dysfunction, which has no basis in fact--no organic correlate. Now, clearly, there are some cases of children with organic problems where this treatment may well be important. But in the overwhelming majority of cases the problems are a reflection of the current state of our crowded schools, overburdened teachers and families, and other social problems rather than something wrong with the kids. Imagine, as biochemical psychiatry is providing more and more information on the biochemical basis of mental states, the construction of a gene that will help to produce a substance in human cells which will change the mental state of individuals. Then, instead of feeding the kids a drug every day, we just do some genetic surgery and it's over.

Don't forget that introducing genes into humans--genetic engineering-- results in permanent changes. There is no way to cut the genes out. It's irreversible. At least, when protests were mounted in certain schools against the drugging of kids, the treatment could be stopped. That's not the case with the genetic solution. There's no going back.

Another example: A current idea, again without scientific foundation, is that aggression is determined by hormone imbalance. Males, it is said, are more aggressive than females because of the hormone testosterone or the absence of presumed female hormones. As a result, patients in mental institutions deemed aggressive are treated with the presumed female hormones.[19] But recently it has been discovered that there are genes in bacteria that will break down testosterone. Wouldn't it be a simpler, less costly approach to introduce such genes (in a functional state) into the "aggressive" patient. Maybe even social protest can be prevented that way. But what are the sources of aggression in this society? Isn't it possible that rather than hormone imbalance, it is social and economic imbalance--unemployment, racism, etc.--that spurs many people to "aggressive" behavior? And, while we're on the subject, would such genetic surgery be used on those in leadership positions in the society responsible for such atrocities as the Indochina war?

Similar approaches could be used to argue for gene therapy on fetuses, infants, or on the workers themselves so that they can work in factories with high vinyl chloride levels. Given the sophistication of the new technologies, a new eugenics era may do even greater damage than the earlier eugenics movement (1900-1930).[20]

CONCLUSION

I would like to add a component to the benefit-risk discussion of recombinant DNA that has, for the most part, been ignored. This component is the risk of human genetic engineering to those without power in this society. Given the present social context, I believe these consequences are inevitable. It is not just the particular evils and damage to individuals I have mentioned in my scenarios that concern me. The dramatic developments in this field, and the publicity they have received and will continue to receive, is already reinforcing the focus on the genetic fix. On the one hand, an atmosphere is being generated in which a variety of genetic approaches to social problems is accepted. And, as a corollary, social, political, and economic changes are deemphasized. The priorities of the society cannot be allowed to be dictated by the technocrats and their technology. On the contrary, technologies must be developed only after social decisions that they are wanted and needed.

On this basis, I believe we should seriously consider whether recombinant DNA research should be pursued at all.

REFERENCES*

1. Shapiro, J., L. Eron, and J. Beckwith. 1969. Letter. Nature 224: 1337.
2. Bergmann, F. H. 1976. A bank of mammalian DNA fragments. Science 194:1226.
3. Pollack, R. 1976. Tumors and embryogenesis. Science 194:1272.
4. T. Friedmann and R. Roblin. 1972. Gene therapy for human genetic disease. Science 175:949.
5. King, J., J. Beckwith, and L. Miller. 1976. Genetic screening: pitfalls. The Science Teacher 43(5):15.

*Many resources on the subjects covered in this article can be obtained from Science for the People, 897 Main Street, Cambridge, Massachusetts 02139.

246

6.  Ausubel, F., J. Beckwith, and F. Janssen. 1974. The politics of genetic engineering: who decides who's defective. Psychol. Today 8(1):30.

7.  Ann Arbor Science for the People. *Biology as a Social Weapon*. Burgess Publishing Company, Minneapolis. In press.

8.  Block, N., and G. Dworkin. 1976. *The IQ Controversy*. Pantheon, New York.

9.  Beckwith, J., and J. King. The XYY syndrome: a dangerous myth. New Sci., November 14, 1974, p. 474. Borgaonkar, D., and S. Shah. 1974. The XYY chromosome, male--or syndrome. Prog. Med. Genet. 10:135.

10. Hook, E. B. 1973. Behavioral implications of the human XYY genotype. Science 179:139.

11. Wilson, E. O. Human decency is animal. New York Times Sunday Magazine, October 12, 1975, p. 38.

12. Powledge, T. Can genetic screening prevent occupational disease? New Sci., September 2, 1976, p. 486.

13. Timnick, L. Next job application may include your genotype too. Houston Chronicle, April 4, 1975.

14. Kilian, D. J., P. J. Picciano, and C. B. Jacobson. 1975. Industrial monitoring--a cytogenetic approach. Ann. N.Y. Acad. Sci. 269: 4. Abstract from the National Cancer Institute, Carcinogenesis Program, 4th Annual Collaborative Conference, February 22-26, 1976.

15. Goodman, E. A genetic cop-out. The Boston Globe, June 15, 1976, p. 37.

16. Ryan, W. *Blaming the Victim*. 1971. Random House, New York.

17. Yao, M. Scientists split on DNA research. The Michigan Daily, March 4, 1976, p. 1.

18. Schrag, P., and D. Divoky. 1975. *The Myth of the Hyperactive Child*. Pantheon, New York.

19. Money, J. 1970. Use of an androgen depleting hormone in the treatment of male sex offenders. J. Sex. Res. 6:165. Blumer, D., and C. Migeon. 1975. Hormone and hormonal agents in the treatment of aggression. J. Nerv. Ment. Dis. 160:127.

20. Beckwith, J. 1976. Social and political uses of genetics in the U.S.: past and present. Ann. N.Y. Acad. Sci. 265:47.

DISCUSSION

RONALD CAPE, Cetus Corporation: I am President of Cetus Corporation, referred to several times by various speakers. We are a tiny company in Berkeley, California. Our funds come exclusively from people who have invested in our company, and we receive no dollars from any government agency.

We are so small, in fact, that when we began in business, with dedication, as Dr. Signer pointed out, to applying molecular biology to industrial problems, someone quite rightly pointed out that we didn't know what we didn't know. And in fact in *Business Week* a couple of weeks ago it was stated, "One problem that Cetus did not understand at first was the difficulty of scaling up production from test tubes to 50,000-gallon automated process units of the kind used by drug companies. This is in the manufacture, five years ago, of antibiotics. Now, I am inexperienced in this kind of public forum, but maybe you can understand why I am flabbergasted to hear Mrs. Simring of Friends of the Earth referring to that and telling you that --I think I heard her correctly--Cetus is going to scale recombinant DNA organisms to 50,000 gallons. I think that kind of thing is a disservice to Mrs. Simring's very deeply felt concerns.

To Dr. Signer's point, I would respond that, yes, we are excited, and yes, we do think that if favorable outcomes are realized we may see a new industry. That is Dr. Baltimore's phrase, and we hope to be part of it. After all, it has been almost impossible to enter the pharmaceutical industry, made up of huge companies only. It is virtually impossible for a new, tiny company. Well, we hope maybe we are excited enough and good enough to change that.

I am further astonished that when we make the statement that we are excited about the fact that there haven't been any applications of molecular biology and we hope to do something about it, Dr. Signer concludes, which seems to me a non sequitur, that we have a solution for which there is no problem.

There are two other things I want to say just briefly. There is no monolithic industrial organization. To that I say thank God, but unfortunately that creates a very real problem in dialoguing with the public and with government. I very much would like to see somebody take the initiative so that companies who want to do the right thing by this issue can in fact find themselves in a forum in which they are having a meaningful dialogue toward solving these problems.

One final word. I think it is very bad for the proponents of recombinant DNA technology to dismiss far-out fears as unlikely due to what they now cite today as technological barriers. First of all, history tells us otherwise. Just recall how recently the difficulty of expressing in bacteria the genes of higher organisms was cited as the reason not to worry about it. I think discussion of all these fears and what to do about them is appropriate.

My final comment is that I am very upset at the polarization and the politicization that I see. I even see delight at the polarization at this meeting. I think we will all be better served in the future if the dialogue continues with more light and less heat.

BERNARD DAVIS: A conspicuous and fundamental feature of Dr. Beckwith's analysis is the emphasis on dichotomies. If you are not with us, you are against us. If you believe that there may be genetic components in behavior, then you are against any efforts to do anything about social components of behavior or of social problems. I don't know anybody interested in behavioral genetics who denies that there are large social components to all social problems. I don't know any basis for implying that people who are interested in also looking at conceivable genetic components of individual problems or individual differences or social problems also are against efforts to understand and to improve the social aspects of whatever the problems may be.

I recall that you stated that it is a disgrace that this government is funding such shoddy research as that in sociobiology. Are you opposed to the use of government funds to support any research concerned with genetic aspects of human behavior? And if your answer is no you are not so opposed, would you mind explaining what you would concede might be desirable?

JONATHAN BECKWITH: The answer is no, and Dr. Davis, you have totally misrepresented my position. I suggest that you read my paper, which I will send you a copy of so that maybe you can listen better.

LEE ROZNER, National Institutes of Health: For the past approximately ten years I have enjoyed at NIH tremendous latitude and freedom of inquiry, perhaps even greater than it exists at so-called academic institutions. All this makes me wonder how the recent successes of science are affecting science itself. I can suggest a few examples.

One point is that as a new technique is invented or is developed in science, a tremendous number of people will rush to use that technique for whatever they can use it for, and the effect of that is autocatalytic. That is, the more people working with restriction enzymes, the more is learned about restriction enzymes, the more they will be used and the more experiments will be designed so that they can be used. So I think that one of the things that is happening today is that there is a diversion of scientific effort into other areas than have previously been considered very important, and a tremendous flow of scientific talent into profitable areas.

One of the things which has puzzled me greatly in this entire meeting is the fact that time after time very noted scientists whom I have had the privilege to be associated with get up and talk about the potential practical uses of the new technologies. These are scientists who I used to hear getting up and talking about how excited they were about going into the lab and doing experiments. These are scientists who would normally have said what we are interested in is basic research. We are interested in inquiring about nature, not enslaving nature. And now I hear them getting up and saying, oh, we are going to do this and we are going to do that. So I think that there is a kind of hypocrisy which scientists are now getting themselves trapped into. It is interesting that the public is suggesting that we do basic research and find out what we are talking about before we start applying it.

One further question occurs to me because of the nature of my work. What happens when there are two ways of going about constructing a particular strain of *E. coli*? And one way is the genetic engineering way, or the DNA recombinant technology way, and the other way is by more traditional methods. Suppose there are two contestants for a grant who wanted to perform a particular project. For example, the very interesting question in molecular biology is replication of the *E. coli* chromosome. Where is it initiated; what is the nature of the DNA from which replication originates in *E. coli*? It can be approached by DNA technology; it can be approached by more subtle, perhaps more difficult, more time-consuming methods of traditional molecular biology. Which one of those approaches is going to get the grant? That means that there is going to be a pressure now on people who are doing research to do the faster research.

We should also keep in mind the effect of science upon scientists. I think that if we were to spend more time and allow more passage of time there will be other ideas which will occur to us, other areas which have not at all been considered. We owe it to ourselves and the public to postpone our activity in this area so that we have an opportunity to really think and really learn what it is that we are dealing with.

STANLEY COHEN: During the past two days, one of the points that has come out at this meeting is the fact that a number of people here are not concerned primarily about the safety issues of recombinant DNA research, but are concerned about what might be done with scientific knowledge in this area. This is an important issue. However, I would submit that knowledge itself is not immoral, but what society chooses to do with that knowledge may sometimes be immoral.

The point of testosterone came up in the discussion a few minutes ago. Clearly society currently has the knowledge to eliminate testosterone production in aggressive individuals without resorting to recombinant DNA techniques, but castration of such individuals is not a socially acceptable practice.

Similarly, methods to accomplish the eugenic goals that Dr. Beckwith finds abhorrent and that I find equally abhorrent also exist at present, but society has not applied these methods because society finds them similarly abhorrent. And in the past when a totalitarian society began to apply those methods the rest of the world found it abhorrent.

The point that I want to make is that we must be careful about preventing the acquisition of knowledge that can be used in ways that are beneficial to society because we are afraid that society may not be able to deal morally with that knowledge. What can be said of recombinant DNA research can be said of virtually all knowledge. Some participants of this Forum seem to have a special fear of genetic knowledge. However, knowledge about genetics didn't begin with recombinant DNA research. It goes back to Mendel for classical genetics and to Watson and Crick for molecular genetics. And one can regard this as a continuum.

I would urge those who are as concerned as I am about assuring moral applications of knowledge to address themselves to what is the real issue as I see it, and that is what is done with knowledge by society, not the knowledge itself.

ROBERT MURRAY, Howard University College of Medicine:  I am a member of the Panel for Inquiry and haven't spoken until now because I didn't see any need to.  And in that role I want to correct or at least have amplified some things that Dr. Baltimore, I think, passed over very quickly in his presentation of the utility of genetic therapy.

He didn't mention at all what kind of vector he proposed to use in such therapy.  If one were to use the virus, there are the hazards of viral particle-shedding carrying such genes getting into other people, and perhaps introducing genetic material where none would be desired, in members of the immediate family, etc.

How can we be certain that genes will not invade the germ line? Even if you put such genes into the bone marrow, such cells are turning over genetic material not incorporated into the chromosome but released to the environment of the host, that is, the human individual, and therefore the possibility of such genetic material getting into the germ line, and then being passed on to offspring who themselves would not be diseased but would likely be carriers of such genes and cause a genetic imbalance.

The fact of proper insertion and regulation is not insignificant. It is critical.  We know that gene position is very important and moving a normal gene to the wrong place in the chromosome can cause lots of difficulty in the organism.

Screening is not hugely expensive.  If you were to present cost-benefit analysis as has been done in screening of the proposed engineering technique balanced against that of screening today, which is likely to come cheaper rather than more expensive, I think screening would come out far ahead.  Moreover, screening technology is reversible. It offers the individual a chance to change his or her mind.

The point was made by Dr. Beckwith that once the gene is in there, how do you get it out, how do you stop its functioning if it starts doing bad things.  It is unfortunate that he chose sickle-cell anemia to illustrate his point.  Sickle-cell anemia is a disease which is compatible with reasonable longevity now, which is increasing with good medical technology and proposed modifications of a variety of drug therapies which can produce what may be a life compatible with normal existence.  And when we have euthenic methods of dealing with such diseases I think we should opt for those rather than genetic, even when that technology may be developed.  Perhaps he should have used Tay-Sachs as a disorder, one that he is probably much more familiar with for a variety of reasons.

His comments about humane motives, I think, are important.  I don't think scientists, and I include myself as a scientist even though I am primarily a clinical geneticist, have any better understanding of what is humane than nonscientists.  Perhaps the nonscientists understand it much better.  I think it is that scientists are members of a public, and we are all members of the public depending on how you look upon it.  Therefore I think the business of defining the public is a matter of where you belong or where you think you belong, and that scientists should talk about joining the public rather than the public joining them.

I hope you don't take too much offense to this, but scientists have been characterized by others more intelligent than I as "kept" people and have been likened to those members of the oldest profession. In other words, we work by virtue of the fact that the public has confidence in our ability to add to knowledge and to do benefit for them. They provide us with money which we do not necessarily earn by the sweat of our brow, through their taxes, and for which we give back to them benefits which we hope will come out of our research. Of course, much research does not give any direct benefit. Therefore, I think scientists should join the public in trying to determine the direction for future research and benefits for those people who may be at risk from some of the work that we do.

DAVID BALTIMORE: I want to thank you for a least taking seriously what I was saying, because I think it is important in these discussions that we develop a dialogue in which we listen to each other.

I think the technical issues that you raised are not a problem. I don't specifically want to go designing molecules right here, but we can envision vectors which are not themselves infectious viruses, and so they will stay in the cells in which you put them, and there is no known way for a gene to go from a somatic cell back into a germ cell, at least to my knowledge.

The question of proper regulation, the proper synthesis of molecules is not a simple problem. But I have a suspicion, and it is just a suspicion, that it will turn out to be a little simpler than one might think, and that the complicated question of position effects and this and that may in fact be taken care of by where the vectors happen to integrate.

The question of screening depends on what you are screening. Things which occur at relatively high frequency in a given population can be screened effectively. Things that occur at extremely low frequency in populations are much harder to get at, and that is why in general most screening, as I understand it, is done only afterward. But I could be wrong. It is certainly not an area I know. As you pointed out, I have my own perspective on the world which is clearly different than yours.

The question of what is humane, I could not agree with you more, is certainly not an issue for me to decide or for you to decide, but is an issue for us all to decide together, and that is ultimately, in fact, what the definition of humane is.

I want to make one other comment that comes out of what Dr. Rozner said. I think that there has been a terrible degradation of science as an enterprise through this discussion, which started now three or four years ago. Science is a matter of the excitement that one gets from the results in the laboratory. And when I grew up in science there was no question that that was good. The reason for that was the assumption that if a scientist was learning truth, the truth was beneficial to the society. We have heard now many questions raised about whether truth is in fact beneficial to the society, because if

it is then we can go right back where we started from. We don't have
to justify what we are doing on the basis of crops or famine or anything else. We can simply say we are providing truth to the society,
and that is what the society wants from us. And from that truth historically has flowed all sorts of things, good things and bad things.
And the things that are good that have flowed from truth, in my world,
are generally medical. So when somebody comes to me, as they did in
Michigan, as people have now come to many of us on many occasions and
said, what good is what you do, my answer has to be in the realm of
medicine because that is what history tells us.

When Pasteur went in to find out what was going on in wine and beer,
I don't know that he knew he would find the basis of infectious disease, but he did. And he didn't have to justify the search for truth
in effectively a mundane area on the basis of what would come out of
it. I don't think we really need to do that justification today
again, and as I think it was Dr. Murray who pointed out that in less
sophisticated circles than this, the idea that truth is good and that
scientific knowledge and scientific progress is good receives a much
happier response than it does in the present circumstance here.

So maybe there is some faith left in the idea that knowledge is
beneficial. But if there isn't, then we have to justify it on the
basis of what we can see in the future in very precise terms. That
is what the politicians are asking us to do. That is what everybody
is asking us to do, and we are doing, I think, about as well as we can
do with it. But it doesn't look very good because it can't promise
specifics. It can only promise generalities, and it is up to you
all to decide whether those generalities are really going to be there
or not.

ETHAN SIGNER: May I remind people that recombinant DNA is not a search
for truth, it is not a search for knowledge, it is a technique.

HESSIE TAFT, Educational Testing Service, Princeton, N.J.: I am a chemist
and I am also a member of the review board in Princeton whose function
parallels that of Cambridge in studying the question of recombinant
DNA research at Princeton University. That committee is also interested in making an intelligent and informed recommendation.

My point today is rather small, but I think important. I would
like to say that I am quite amazed at the way Dr. Valentine presented
scientific data to an audience that should be or is expected to be
highly critical. I may be somewhat misquoting you, but I jotted down
what I thought I heard, which was that Mother Nature, not recombinant
DNA, has built an incredible energy barrier to prevent the overproduction of ammonium ions during nitrogen fixation. And then you proceeded
to show us a slide of data consisting of four microbes studied, one
of which didn't seem to fall into line. That one was pointed out as
needing further verification. Well, may I suggest that I am not quite
willing to accept this as convincing data, and it seems to be the kind
of thing that could lead one typically to the criticism--it leaves

oneself open, I think, to the criticism that many people are accused of in this area of rushing into things without proper scientific background.

I am reminded of the old story of the mathematician asking the physicist for proof of the following conjecture: All odd numbers are prime, true or false? And the physicist says yes, one is a prime, three is a prime, five is a prime, seven is a prime, nine is an experimental error, eleven is a prime, and therefore the conjecture is true. I think we had better be careful before we run into the same kind of mistake now.

RAYMOND VALENTINE: Well, I am sorry I have offended you with this data. I think we should have really had more data to criticize this way.

In using the slide that you refer to, I was trying to be nice to one of my most prestigious colleagues. Actually from his laboratory came two of those values. One was the twenty-nine value and the other was four. His own laboratory has since extracted the enzyme from the low value and shown that at the biochemical level, mechanistically, it uses far more energy by a factor of three to four than he obtained in his living system.

So I think this is the real point. I put that value there to give you some balance in this figure. It is not overwhelming. I said later on, in discussing the blue-green algae I had on the slide, that more data is needed.

MEREDETH TURSHEN, Oil, Chemical and Atomic Workers International Union: I would like to be brief, but if I understand Dr. Baltimore correctly we are supposed to footnote our remarks. I would like to address myself in some sort of positive and constructive way because I have been challenged to do so by people who have come up and spoken to me at the various breaks, to outline--very briefly--what we consider, those of us who are challenging recombinant DNA research, to be a positive alternative to this research.

Many of us do advocate not just a moratorium, but a cessation of this type of research. Now, in the area of the high-yielding varieties of rice and wheat, I need not go back over all of the research that has been done--footnote, United Nations Research Institute for Social Development; Keith Griffin, Professor of Economics, Oxford University; Ingrid Palmer, Professor of Economics, Hull University, etc.--which has studied the results of the so-called Green Revolution, and found what is now known as its black side, referred to very briefly by Ethan Signer.

I would just like to point out that study after study in country after country beginning with India but going on through Southeast Asia has documented that the poor are absolutely poorer as a result of the introduction of this technology, that they are eating absolutely less than they were ten years ago and fifteen years ago, not to mention the fact that they are all eating much less than they were twenty-five, thirty, and one hundred years ago.

I think that it is not and never has been a problem of food production in the world.  It has been and always will be, until we
resolve it politically, a question of food distribution in the world.
And those of us who are opposed to the kind of research we were
shown--nitrogen fixation--wish to see programs introduced, beginning
in this country where there is still hunger, food distribution, income
guarantees, food stamps not based on a needs test, but a guaranteed
adequate, balanced diet to every American.  And there might just be
spin-offs to the diabetics who are overweight.

# AN OVERVIEW OF THE ROLE OF NIH AND OTHER FEDERAL AGENCIES IN THE CONDUCT OF RESEARCH WITH RECOMBINANT DNA

Donald S. Fredrickson

*Director, National Institutes of Health*

I am very grateful for the opportunity to summarize something about the government process in this matter which you have been discussing throughout this Forum.

Governments in general, and the federal government in particular, entered the matter of recombinant DNA research several years ago when, after Asilomar as you recall, one of the recommendations of the scientists was that the NIH form a committee that might begin to set up guidelines to establish strict conditions for the conduct of research that involved the use and production of recombinant DNA molecules. These guidelines were developed by a recombinant DNA committee, and after extensive scientific and public review the NIH released them on June 23, 1976.

The provisions were designed to afford protection with a wide margin of safety to workers and to the environment. The NIH guidelines were published in the *Federal Register* on July 7, 1976, for public comment.

In September the NIH also filed a draft environmental impact statement on the guidelines for public comment, and the final NIH environmental impact statement we expect to be published shortly. As many of you are aware, in August 1976, a volume was published by the NIH that contains the transcript of a public hearing held on the guidelines as well as all correspondence received by my office on this matter prior to the release of the guidelines in June. And there will be a subsequent publication of all correspondence and many other related documents to continue this complete public record of government action in regard to recombinant DNA research.

By the time the environmental impact statement had been issued and the guidelines released, it was already apparent that the international community of science had come to agreement that recombinant DNA techniques should be used only with a common set of standards across the world. The question was how to bring this about. And as matters of this sort are often settled, first the boundaries of activity were restricted to those maximum ones in which the law can be operable across a population, and hence most countries settled down to attempt to enter this second phase for themselves before seeking further international comity and conformity with particular standards.

At the time the NIH guidelines were released there was convened by the Secretary of Health, Education, and Welfare an Interagency Committee on Recombinant DNA Research. The committee was formed with the approval of the President, and at the Secretary's request I have served as chairman.

Now, this Interagency Committee is composed of representatives of the federal departments and agencies that do several things. One component is made up of those agencies that support or conduct recombinant DNA research, or may do so in the future. Another group includes the representatives of all the federal departments and agencies that have present or possibly potential regulatory authority in this area. And to these are added a number of other departments, such as the State Department, the Department of Justice, and others that have particular interest in the general aspects of the committee's affairs. There are approximately twenty-five member agencies that make up this committee.

The mandate of the Interagency Committee is to review the nature and scope, particularly of the federal activities, relating to recombinant DNA research. Second, the committee was directed to determine the extent to which the NIH guidelines may currently be applied to research in both the public and the private sectors. It was to recommend, if appropriate, legislative or executive actions necessary to ensure compliance with the standards set for this research, and to provide for the full communication and necessary exchange of information on recombinant DNA research programs and activities throughout the federal sector.

The Interagency Committee held its first full meeting last November, and during that month it had a second meeting. The first of those meetings was held on November 4 and was devoted to a review of the development of the NIH guidelines for research involving recombinant DNA molecules. At the same meeting the committee also reviewed *in extenso* international activities relative to this same matter. I will not repeat that review, because I understand you are to have a report of a workshop which will summarize it for you in much greater detail than I can now. But the committee was fully aware of activities relative to this matter not only in this country but abroad.

At the meeting of the committee held on November 23, the federal research agencies then discussed their activities and possible roles in the implementation of the NIH guidelines. All of the research agencies endorsed the NIH guidelines to cover the recombinant DNA research that they conducted or funded. Three agencies of the federal government are now supporting research that involves the use of these techniques, the NIH, the National Science Foundation, and the Department of Agriculture. The Department of Defense, NASA, and the Energy Research and Development Administration are not at present conducting such research, but agreed to use the NIH guidelines to govern future research should they undertake it.

In that November 23 meeting the federal regulatory agencies also reported on their regulatory functions. Following that lengthy review a special subcommittee was set up to analyze the relevant statutory authorities for the possible regulation of recombinant DNA research. All regulatory agencies were represented on that subcommittee, and their representatives were assisted by attorneys from their offices of general counsel.

The subcommittee was charged to find out whether existing legislative authority would permit the regulation of all recombinant DNA research in the United States, whether it was funded by the government or not, and to seek out whether or not those existing legislative authorities would include at least the following requirements perceived by the committee to be important: review of such research before it is undertaken by an institutional biohazards committee; compliance with physical and biological standards and prohibitions in the NIH guidelines; registration of such research in a national registry; and enforcement of the above requirements through monitoring, inspection, and some sanctions.

It was the conclusion of this subcommittee after extensive review that present law permits imposition of some of the desired requirements on much recombinant DNA laboratory research, but no single legal authority or combination of them currently exists that would clearly cover all research or other uses of recombinant DNA techniques and meet all the other requirements. The committee examined, first of all, the Occupational Safety and Health Act, and found that while OSHA has broad authority it has limited access to many of the laboratories, and it does not cover self-employed persons. The Environmental Protection Agency under the Toxic Substances Control Act is directed to control chemicals that may present an unreasonable risk of injury to the health or the environment. The subcommittee found that probably most recombinant DNA molecules could come under the definition of chemicals; however, Section 5 of the Toxic Substances Act explicitly exempts registration of chemical substances used in small quantities for the purposes of scientific experimentation or analysis. The latter exemption represents the most serious deficiency in the authorities of that act for the purposes of regulating the use and production of recombinant DNA molecules.

The Hazardous Materials Transportation Act was also examined. It gives the Department of Transportation and the Center for Disease Control in Atlanta considerable authority over interstate shipment of hazardous materials but, indeed, there were many aspects of this act which are wanting in regard to regulation of recombinant DNA research.

The Environmental Defense Fund, in November of 1976 petitioned the DHEW to regulate recombinant DNA research under Section 361 of the Public Health Service Act, and this petition was filed with the Interagency Committee for its consideration. Under this section the authorities are restricted to organisms that are communicable and cause human disease. To use Section 361 for regulatory authority one would have to assume that recombinant DNA research may cause human diseases and that these may be communicable. Further, Section 361 does not apply to plants or animals or the general environment. It was the conclusion of the committee that Section 361 lacks the requisite authorities.

The same is true of Section 353 of the Public Health Service Act. This applies to clinical laboratories, but it is not considered to be applicable to research laboratories.

Many other authorities, particularly of the EPA and of other agencies including the Food and Drug Administration, were examined, as were the powers of the Department of Agriculture, whose authorities were found applicable solely to nonhuman life and plants.

In summary, the Interagency Committee concluded that no single authority or combination of authorities currently exists that could clearly reach all recombinant DNA research in a manner that the committee deemed was appropriate. It was agreed that regulatory actions could be taken under existing authorities, but that they would be in considerable jeopardy of legal challenge.

The full committee then adopted the report of the subcommittee, agreeing with its conclusion about existing authority. It then turned its attention to examining possible new legislation. In considering elements for new legislation the committee reviewed federal, state, and local actions and activities that bear on the regulation of DNA research. In addition to activities in municipalities such as Cambridge, it received a report from the New York State Attorney General's Environmental Health Bureau for State Regulation, which made certain recommendations for regulation in New York State. The committee was aware of the hearings in the California legislature, and it also was able to examine legislation now submitted to the Congress, specifically Senate Bill 621, the DNA Research Act of 1977, introduced by Senator Dale Bumpers, and the companion measure introduced into the House by Mr. Ottinger.

The committee also had available to it comments elicited by its various members from a number of persons whose opinions were sought concerning questions relative to legislation. These sources included agricultural scientists, biomedical scientists, environmentalists, and leaders from labor unions and private industry. In the light of this background the committee has been considering in its most recent meetings what should be the elements of new legislation that might cover the regulation of the use and production of recombinant DNA molecules. It has had to consider issues of definition, the question of registration of all activities, and the question of whether licensure might be an effective part of a regulatory process, and it has dealt strongly with aspects of interagency cooperation. It also has had to deal with the difference or the distinction between research and commercial use of recombinant DNA products, particularly because many commercial aspects are clearly covered by existing legislation or authority invested in the Environmental Protection Agency and the Food and Drug Administration. It has also had to contend with the fact that the NIH is not a regulatory agency and that it has no intention of becoming one, and that it would be inappropriate for NIH to assume inspection and enforcement authorities when it has participated in standards setting.

The Interagency Committee meets again tomorrow. We expect and hope that it may produce an interim report dealing with some recommendations with respect to legislation within a week. Then the committee will turn to other agenda relative to this problem, and at some point will probably self-destruct when it has carried out fully the terms of its mandate.

In brief, there is a strong and active focus within the executive branch to formulate recommendations to help set federal standards, which I think to be very much needed, with regard to the regulation of the use and production of recombinant DNA molecules. The task has not been simple. The committee has recognized its responsibility to protect fully the public interest. It recognizes that recombinant DNA activities can pose some threat to workers, to the general population, to the environment, and also to a creative and responsible scientific apparatus. Thus, the task of recommending appropriate, effective, and reasonable legislation for regulation of this activity is a matter of very grave concern.

# DISCUSSION

NORTON ZINDER, Rockefeller University: I would like to support, and I am surprised that I am going to do so, the idea of having legislation, federal legislation, with regard to recombinant DNA research. The proliferation of local option with different guidelines in different states and different cities can only lead to a situation of chaos, confusion, and ultimately to hypocrisy amongst the scientists involved. I strongly plead that the government move ahead on this as rapidly as possible.

FREDRICKSON: Thank you, Dr. Zinder.

AL PLUMMER, retired civil engineer: I am neither for nor against rapid research in recombinant DNA. I am here to learn what the facts are so as a private citizen I can choose sides when it becomes appropriate. I have listened to 85 percent of the discussion, and so far I have not been able to identify who in the federal government is responsible for bringing together in one comprehensive document all the history, facts, alternative pathways, along with calculations, as best they can be made, of benefits and costs and the environmental impacts of this problem. In other words, who is doing the planning that will point out where we are going so that we as private citizens can make intelligent decisions? Is there a group planning what kind of a program would be appropriate for the nation as a whole? I understand you are dealing with regulation and setting standards, and that is fine, but it doesn't really attack the problem of where we are going with this. Is it good? Is it bad? What are the problems? Can you tell me who is going to come up with a document?

FREDRICKSON: Yes, Mr. Plummer. First of all, several documents have already been issued which may be helpful to you. I referred to two of them. One is the NIH preparation of the history relative to its guidelines. The second is the environmental impact statement issued relative to its guidelines. The Interagency Committee now contains all the elements of the federal supporters and conductors of this research, and probably they are responsible for at least 90 percent of the research that is doubtless going on at the present time in this country. They will be reporting to the Secretary of Health, Education, and Welfare, in whose office now, as this matter ascends higher up in the Department, will be the next focus for disseminating and developing some of the considerations that you represent. A third focus will open up next week when the Congress of the United States will, I think, have the first of a number of hearings on this whole matter of recombinant DNA research. I believe that there are several committee hearings that are scheduled or are about to be, which will deal not only with the matter of legislation, but also the general aspects of the recombinant DNA research. Finally, we have been this week, and I expect to return to, the Appropriations Committee in the House, and the Senate next week, where we have also been answering a number of questions of the kind that you have posed.

PLUMMER: Well, my basic statement is that it is fragmented and it isn't pulled together in such a way that we can quietly analyze it and come to conclusions. As a result I see in this meeting that the opponents and proponents are polarizing and that will lead to emotional situations, and it will get more and more difficult to resolve unless we get the facts all laid out.

DAVID O. KRASSIK, Engineering and Applied Science, UCLA: I am interested in learning more about how one has established the adequacy of the current NIH guidelines. My background is not biology or biochemistry, and I have been listening to try to keep the words *vector*, *phage*, and so forth apart in my mind.

I did go to the containment workshop last night and there tried to find out whether there exist documents that would give details on the efficacy of physical containment and biological containment, but I was told no. I must confess, I was a little surprised at what seemed to be the relative ineffectiveness of P1 to P3 containment, assuming that there is a risk, and I have to rely on my medical and biological colleagues to tell me that.

With regard to the biological containment, again, one hears numbers of large factors, but again, there are uncertainties. So as I listened I found in my own mind no way of knowing as a result of these few days how the guidelines were arrived at, how the kinds of questions raised by Dr. Sinsheimer here at the Forum yesterday and previously have been dealt with or are being dealt with in deciding that these guidelines are adequate, that they are not too strict or not strict enough.

FREDRICKSON: Have you had opportunity to read the NIH guidelines, their appendixes, and all of the comments relative to them?

KRASSIK: Yes, I have, but with my limited background I could only digest part of it.

FREDRICKSON: If you will give us your name at the NIH one of the documents that will be very helpful to you is the revised or final environmental impact draft statement, which has addressed in detail comments of the kind and questions of the kind that you have raised. The development of both the guidelines in their final form and the environmental impact statement have involved an exchange of correspondence and a full attention to a wide range of public comment, each of which has been addressed in the revised document.

FRANCINE SIMRING, Friends of the Earth: I would like to congratulate the NIH and Dr. Fredrickson on the wonderful job they have done of disseminating materials, transcripts, and Xeroxes to all interested parties. And in the interest of expanding the accuracy, I want to make three short additions to what was said by Dr. Fredrickson.

You mentioned, I believe, that all correspondence was included in the yellow volume of August. I believe we would have to make that "some" correspondence in the interest of accuracy.

You mentioned that the nations settled down by themselves to do their guidelines. A few did, but for the most part in the correspondence that I read, many nations wrote to state they are looking to the United States for leadership, and will follow the U.S. guidelines when they are published. In the light of this afternoon's press conference, I think that is particularly important.

The last point that I would like to make is that Dr. Fredrickson mentioned that the Interagency Committee listed registration of such research with a national registry. However, there is a parenthetical opening for industry that reads as follows: "Subject to appropriate safeguards to protect proprietary interests," which means that they might not have to register their projects.

FREDRICKSON: Thank you, Mrs. Simring. I am glad to meet you even at this distance, and I hope to close the gap between us.

Indeed, the volume that I referred to does refer only to correspondence relative to the guidelines. Many of the subsequent letters we have received will have a broader base because more action has occurred since that time. There will be another issuance. I know that some of your correspondence will also appear there.

With respect to extension of the U.S. guidelines, it is true that there are other countries that are using them, as well as the United Kingdom guidelines. You will hear more about that, I am sure, in the final description.

With respect to the matter of registration and the issue of proprietary information, this is certainly one matter which the Interagency Committee is discussing and will grapple with completely, you can be sure.

AUDIENCE: Dr. Fredrickson, like many people I am beginning to share a mania against federal intrusion into so many aspects, and I think it is rare and unique for the American scientific community to actually invite a federal incursion. Yet, with so many recent experiences, occupational safety and health and what have you, it has been proven that the federal government is probably the least adept. I am happy to see that you are working with the Hill on legislation, but you just mentioned that the NIH doesn't have or want enforcement authority regarding this work. Who would the legislation give it to? HEW? Federal bureaucrats? Who is going to monitor it? I assume that there will be legislation, but I hate to see an element of control removed from the scientific community, and I wonder who they will award it to.

FREDRICKSON: I am not sure to whom it will be awarded either. But you can be sure that this is a matter which the committee is now actively considering and will deal with in its report.

# THE ECONOMIC IMPLICATIONS OF REGULATION BY EXPERTISE: THE GUIDELINES FOR RECOMBINANT DNA RESEARCH*

Roger G. Noll

*Professor of Economics, California Institute of Technology*

Paul A. Thomas

*Graduate Student in Social Science, California Institute of Technology*

The debate over recombinant DNA research raises a number of important issues of public policy. Receiving most attention has been the direct

*Part of the cost of preparing this paper was financed by a grant from the National Science Foundation Program of Research Applied to National Needs, Grant APR 75-16566A01, and by a National Science Foundation graduate fellowship.

question about the social value of the research, considering its potential benefits and risks. Equally important, but receiving somewhat less attention, are a series of more general issues that, while illustrated by the debate over recombinant DNA research, are likely to recur in other contexts with increasing frequency. First, to what extent can and should society constrain and direct scientific research? Second, in making decisions that require the use of highly technical information that is possessed by a very restricted group, to what extent can society make decisions that are technically informed without in the process delegating the authority to make nontechnical judgments and evaluations to an unrepresentative technical elite?

Although the guidelines issued by the National Institutes of Health (NIH) have been subjected to public review and are being supplemented and amended by political jurisdictions ranging from city councils to the U.S. Congress, the essential feature of the approach that has been taken to date to control recombinant DNA research is professional self-regulation. The molecular biologists who do this research have established the grounds for debate. Specifically, most of the discussion has focused on classifying the range of recombinant experiments according to the direct risk they pose to humans and assigning to each class a set of safety rules, ranging from outright bans to good laboratory procedures under normal circumstances. Moreover, for the most part implementation of the guidelines is left to the scientists who are in charge of the research. The NIH guidelines provide no enforcement mechanisms other than the requirement that grants from NIH be given only to institutions that agree to abide by them.

Although government organizations at all levels have attempted to review most of the features of the guidelines, government actions thus far have been primarily to consider enforcement mechanisms that would cause all researchers, including those not supported by NIH, to abide by the guidelines or face stiff penalties. NIH is not a regulatory agency, and has neither the resources nor the mandate to engage in the kind of enforcement activities that are practiced by agencies such as the Food and Drug Administration (FDA) or the Occupational Safety and Health Administration (OSHA). Consequently, the obvious first step for legislative and regulatory authorities is to add teeth to the guidelines. Meanwhile, the underlying conceptual model that molecular biologists initially applied to the problem has remained largely untouched by the process of public review.


REGULATING TECHNICALLY COMPLEX ACTIVITIES

Recombinant DNA research, like so many problems of technical assessment, is a public policy issue and a candidate for regulation for two reasons. First, the federal government provides most of the financial support for molecular biology. Consequently, the public naturally will ask what it is buying, and whether particular lines of research deserve more or less public financial support. Second, the public must bear most of the risks of experimental accidents. Even if an experiment is not financed from the public treasury, citizens still have a stake in the safety practices surrounding a dangerous experiment, since an accident can lead to significant uncompensated losses to persons who play no part in the decision to undertake the experiment and, therefore, whose welfare may not be fully taken into account by whomever makes that decision.

## The Role of the Expert

A necessary input to rational policy decisions about sophisticated new technical developments is an assessment of the procedures and outcomes of the various ways the new technique can be used. Most activities at the frontiers of human knowledge, including recombinant DNA research, are fully understood only by highly trained experts. These experts must be involved in the public decision-making process if policy decisions are to be sensible. The problem for public policymakers is to devise a mechanism for gathering the relevant technical information and checking its authenticity and completeness without at the same time delegating to the experts too many aspects of the decision that do not depend on technical expertise.

The dangers in delegating too much authority to the technical expert are more complex than are generally recognized. In the debate over recombinant DNA research, the delegation problem receiving most attention is the direct stake in terms of financial and professional gains that molecular biologists have in the outcome. Certainly this issue is relevant. The biologists who do this research have years of professional training, substantial financial support, and the prospect of receiving professional awards and prestige hinging on the decision whether DNA recombinant research will be permitted. But this argument can cut both ways. The public's *perception* of the riskiness of the research, not the actual risk, will determine the amount and nature of research that will be allowed. Because of the technical complexity of the problem of assessing the risks, public decision makers are likely to be somewhat uncertain about the technical information that is supplied by the experts, even if in reality the information is accurate and complete. If so, a few unnecessary safety precautions that ease public uncertainty may be a small price for the experts to pay in light of the personal gains to be captured by those who engage in the research. Thus, if the research is fundamentally safe but its safety is difficult to prove, the personal stake that scientists have in the issue may well lead to unnecessarily cautious safety standards as an expedient.

Nevertheless, the public uncertainty over risk assessments by experts is a natural, rational response to the disparities in incentives faced by experts and by the public at large. Experts control the information on which risk assessment is based, and they are likely to be willing to run greater risks than would be acceptable to the public at large. An obvious factor contributing to this difference is the personal stake of the experts in continuing the research that requires their expertise. But there are other factors operating as well.

One of the values of research is the excitement of acquiring new knowledge, regardless of its immediate or prospective usefulness. Whether the specific project is unraveling the genetic code, searching for life on Mars, discovering the essence of physical matter, or comprehending more completely the behavior of complex social systems, the act of expanding the frontiers of human understanding is, to some at least, of considerable interest in its own right. Research is, then, a form of consumer good. Individuals can be expected to differ according to the value they place on increasing human knowledge for its own sake, without considering its practical benefits, just as they differ in their tastes for other purely consumptive activities. Consequently, the costs that people are willing to bear in order to pursue new knowledge will differ from person to person.

People who have chosen to do research on any particular topic are not likely to be representative of society at large in terms of their tastes for that research. First, technical experts understand more of the ramifications of new knowledge, and hence can derive more consumptive value from research than nonexperts. Second, anyone who pursues a particular line of scholarly research does so in part because it seems especially interesting to that person. Molecular biologists are a self-selected group. Far more people have the ability and motivation to become molecular biologists than actually do so; others become physicists, lawyers, and even economists. These decisions reflect individual tastes for particular kinds of knowledge, and it stands to reason that molecular biologists will find genetic experiments more interesting than will people who do other kinds of research or who have selected careers that do not involve research. Third, biologists engaged in hazardous research are also self-selected in terms of attitudes towards risk. Just as individuals exhibit different tastes for consumptive activities and occupations, so, too, do they differ in the amount and type of risk that they are willing to accept. If a particular line of work, whether using recombinant DNA techniques or lumberjacking, is of greater than average risk, people who enter that line of work are likely to be, on average, either more risk-taking, or more optimistic in their beliefs about aspects of the field that are still incompletely understood, than are people in general.

For the preceding reasons, the public at large is likely to be less than fully reassured if a particular group of technical experts claims that their line of work is sufficiently nonthreatening to society to be worth pursuing. What is safe enough to people in one line of work is unlikely to reflect an evaluation of risks and benefits that is representative of the values of other members of society.

An additional problem of self-regulation arises if more than one area of expertise is relevant to the policy decision. If a particular expert group regulates its own activities, it faces the same problem with respect to other groups of experts that society faces with respect to it. If other groups are consulted, the self-regulated group loses autonomy and authority, but if it decides to handle all aspects of the problem itself, it will be likely to make errors of analysis in reaching its conclusions. From society's point of view, the quality of the ultimate decision regulating the activities of experts will obviously be lower if relevant parts of the analysis underpinning the decision are overlooked or flawed, while informational inequities make it difficult to consult the affected experts without inadvertently delegating authority to them.

The discussion so far has produced several reasons why citizens may want public officials to intervene in the self-regulatory activities of a particular technical elite. These arguments can be generalized to a simple proposition. The social desirability of a public policy decision depends upon both the quality of the technical information on which the decision is based and the extent to which the decision is representative of the tastes and values of the affected individuals. In certain arenas of public policy, one can acquire better and more complete technical information on one aspect of the problem only by sacrificing some of the quality of other types of information and/or the representativeness of the outcome.

Measuring the extent to which a particular decision is unrepresentative of the decision that a society would make if all members were fully informed is, of course, impossible, since the hypothesized cause of an unrepresentative procedure is the unavailability to all but the expert of the very information that would be necessary to make the measurement. Nevertheless, the logic of the preceding arguments leads to some qualitative predictions that can be tested. First, activities in which experts are already involved are likely to be regulated less tightly than are activities that have been well defined and considered by the experts and that objectively have equal potential risks and benefits but that have not yet been undertaken. In spite of the fact that precise regulation of ongoing activities is easier to devise because more information is available about it, looser regulations will be applied to areas of ongoing activity, all other things being equal, because experts already will have made personal decisions about and commitments to the ongoing activities. Second, an unrepresentative procedure is prone to overlook entirely or to analyze erroneously issues that call for the use of the tools of another discipline.

## AN EVALUATION OF THE GUIDELINES

The NIH guidelines and the justifications accompanying them appear to exhibit these two characteristics of an unrepresentative outcome. The purpose of this section will be to offer some evidence for this proposition.

### Inconsistencies in the Guidelines

The validity of the first prediction regarding inconsistencies in the guidelines that are related to the pattern of ongoing work remains for the molecular biologists to determine, but to an outsider the results are suspicious. The NIH guidelines contain several examples of either unequal treatment of roughly similar risks, or equal treatment of apparently quite different risks. A few examples illustrate the point.

First, the controls on recombinant experiments involving insect DNA are essentially no more than standard good laboratory procedures, while substantially more stringent controls have been placed on experiments involving DNA from lower vertebrates and higher plants. The rationale for this and other differences in controls according to the species from which DNA is taken is that the less related is the DNA to human genes, the less is the risk to humans. Neither the guidelines nor any biological literature of which we are aware provides support for the proposition that this principle should extend to distinctions between insects and trees. Moreover, risks other than the problem of direct threats to humans should be considered. Humans and other species could be affected indirectly if hybrid cells entered and altered food or disease chains at any point. Thus, the distinction between insect DNA and other species subject to tighter controls seems without any real scientific foundation. What is clear is that *Drosophila* DNA has been used in some of the pioneering efforts in recombinant research, and that one user of it served on the committee that wrote the first draft of the guidelines.

Second, the literature on the comparative properties of different hosts and vectors for recombinant DNA experiments suggests that some are more dangerous than others. The text and appendixes of the guidelines contain several informative comparative analyses of alternative source-host-vector systems. For example, simian virus 40, a virus that is known to cause cancer in animals, is generally regarded as less safe than polyoma virus; furthermore, *B. subtilis*, although less well studied than *E. coli*, is regarded as likely to prove safer than the latter; and lambda bacteriophage, although less manipulable by experimenters, is regarded as likely to prove to be safer than the plasmids that are commonly used as vectors. The general principle involved in these safety judgments is that experiments are likely to be safer if none of the elements involved in affecting the DNA recombination have a known niche in man or a closely related species. Nevertheless, in each of the three cases cited above, the controls proposed in the NIH guidelines do not distinguish between the more and less risky alternatives.

The principal basis for the decision to treat these alternatives equally is that more is understood about the genetics of the more risky alternatives, which is a result of the fact that the more risky alternatives have been more extensively used in experiments in molecular biology. Consequently, research on the characteristics of the alternatives would have to proceed for several years before most of the interesting recombinant experiments involving them could be performed. Thus, the decision to have equal treatment of more and less risky alternatives is primarily one of expediency. Of course, the decision has the unfortunate long-term consequence that it provides no incentive for molecular biologists to develop alternative sources, hosts, and vectors that promise to be safer or to use these alternatives if they are developed.

The value of standards as incentives is illustrated by a recent example. The level of biological containment prescribed for the most dangerous experiments that the guidelines permit could not be achieved at the time the guidelines were originally proposed. Consequently, if some of the most interesting experiments were to be performed, a new host had to be developed that was satisfactory for experimental purposes but weaker than those then in use. Roy Curtiss III and his colleagues at Alabama succeeded in developing a weakened strain of *E. coli* literally within months of the development of a demand for it.

The point of the preceding example is that the guidelines should be regarded as more than a set of controls for existing experiments. They also set up incentives that will affect the future course of research in the field. The failure to provide incentives to develop less risky hosts, vectors, and sources of DNA reduces the chance that they will be developed or that they will be extensively used if they are developed. In short, today's guidelines not only affect the safety of current experiments, they indirectly affect the safety that can be achieved in the future. There is no evidence that this particular long-term effect of the system of controls that NIH has proposed was taken into consideration.

The preceding discussion, of course, must be regarded as raising a series of questions, rather than constituting an indictment of the guidelines. Not being molecular biologists, we cannot be certain of purely technical issues in this highly complex field. With regard to the second prediction--that important issues not within the range of expertise of the perpetrators of the guidelines would be overlooked or dealt with incorrectly--the guidelines do exhibit conformance with expectations.


The Technical Orientation of the Guidelines

The major sins of omission of the guidelines have to do with their purely technical character. Essentially, the guidelines define the laboratory practices, physical layout, and biological containment required for the experiments that are permitted. Numerous other issues that bear crucially on the type and amount of research that will be undertaken, and the attendant hazards that society will face, have been largely overlooked in the debate about recombinant DNA research.

One such omission is a comprehensive analysis of problems of human error. The guidelines specify certain training requirements and laboratory practices (e.g., no pipetting by mouth) for laboratory workers in labs in which recombinant DNA research is taking place. As Paul Berg has observed, the regulations regarding physical containment in facilities at containment levels up to and including P3 are dependent upon the absence of human errors and outright risky shortcuts that are known to take place in laboratories. Consequently, most molecular biologists regard the biological containment regulations as far more important than those regarding physical containment. Even here, however, human error is a distinct possibility, owing to mistakes such as confusing samples or, in the dark of night when no one else is watching, simply taking a shortcut. Undoubtedly human error can never be eliminated; however, the guidelines do not inventory the range of possible human errors that might be especially dangerous, and in so doing miss whatever potential exists for using the guidelines to avoid or ameliorate them.

The debate over recombinant DNA research has also avoided examining the possibility of using budgetary allocations among types of research as a mechanism to alter the direction and safety of recombinant research. The risk to society from recombinant DNA depends on the nature of the research projects carried out in this field, which in turn depends upon budgetary allocations by NIH and the National Science Foundation (NSF), the agencies that provide most of the financial support for molecular biology. Consequently, one mechanism for altering the societal exposure to risky experiments is to allocate more of the budget for research in molecular biology to other types of genetic research and to the safer varieties of recombinant research. In addition, budgetary allocations could be increased for developing safer host-vector systems. Historically, research scholars have been the dominant force in selecting the lines of research to be pursued and, therefore, the way that NIH and NSF spend their research budgets. As a result, taking a more instrumental view of budgetary allocations represents something of a break with tradition that would weaken the influence of molecular biologists in determining the directions of further reserach in their field. At the same time, the use of budgetary incentives may be a more effective mechanism in the long run for reducing the riskiness of research than is direct regulation of the laboratory environment.

Another omission from the discussion about recombinant DNA experiments is a serious, comprehensive treatment of the problems of enforcing the guidelines. The only federal enforcement activities that are contemplated in the guidelines are the threat of the loss of NIH financial support if the guidelines are violated and the creation of an oversight committee to inspect laboratories in which recombinant research is carried out. The committee would include nonbiologists from the local community.

These provisions constitute a very mild enforcement system. The nature of the oversight committee makes suspect its ability to identify violations of the guidelines other than very gross ones. Moreover, the threat that a violation will lead to suspension of all NIH support to a university provides a strong incentive for a basically friendly oversight committee to avoid reporting violations, since members of a university community are unlikely to want to see the university placed in financial jeopardy. And even if a violation is reported, NIH is not likely to carry out the threat to cancel all of its grants to a major research institution without considering the motivation and severity of the violation. Of course this creates opportunities for politically expedient decisions that undermine the guidelines. The source of this problem is that a penalty system that imposes the same punishment, regardless of the offense, does not make much sense. Certainly, a failure to abide by the most stringent containment standards for the most hazardous experiments should be dealt with more stringently than even a premeditated avoidance of some feature of the standards for an experiment with

minimal adverse consequences.  Yet any attempt to make decisions depend
on severity and motivation converts NIH to a judicial authority without
any of the normal legal safeguards of regulatory processes.

Some of the ramifications of the issues not raised by the guidelines
or to a significant degree in the debate about recombinant research are,
of course, not within the existing ambit of authority of NIH.  Without
legislative action, NIH could not make a major change in its budget or
impose a complex penalty scheme on violators of the guidelines.  But it
is reasonable to ask NIH and the community of molecular biologists to
recognize the importance of dealing with these issues, to address them
seriously, and to propose actions that Congress and other governmental
units might take.  The principal issue in the early interventions by
state and local governments, such as California and Cambridge, Massachusetts, has been the problem of enforcement, rather than the adequacy
of the guidelines.  This is a rational public response to the cursory
attention that has been given to enforcement thus far.  Whether the
guidelines can be effectively enforced at reasonable costs, both in
dollars and in loss of freedom of inquiry, remains an open question.

Benefit-Risk Analysis

The primary sin of commission in the debate about recombinant DNA
research and the desirability of the guidelines has been the simplistic
and largely inappropriate use of benefit-risk analysis to evaluate the
research.  In debating the value of their research in terms of benefits
and risks, the molecular biologists have overstepped the bounds of their
technical expertise, with the result that crucial aspects of a valid
benefit-risk analysis are omitted or incorrectly treated in the discussion.  The following are but a few examples to illustrate the point.

The principal benefits that are cited in the discussion about the
value of recombinant DNA research, in addition to the overall contribution to human knowledge that the research will produce, are several
commercial uses of particular kinds of recombinations.  Among the specific possibilities mentioned are the production of insulin, hemoglobin,
and other body chemicals, the development of a cure for viral cancer,
and the creation of plants that use atmospheric nitrogen.  Among the
issues missing from the benefit discussion are:  (1) an assessment of the
probability that any of these possibilities will be commercially attractive, (2) an estimate of the amount of time it will take for knowledge
to be sufficent to make these objectives technically possible, (3) an
estimate of the costs of the research that must be done before society
will know whether commercial use of DNA recombination is worthwhile,
and (4) the design of a comprehensive program of research that would
contribute to the achievement of these public health and agricultural
objectives.  Each of these is essential to calculating the net expected
benefits of the program.  To apply the benefit-risk model to a line of
research requires developing a research program that maximizes the difference between the expected benefits and costs of the activities.  Some

of the necessary component parts of the analysis are developing a calculus for comparing costs and benefits that are separated in time (e.g., How are risks and costs borne by the current generation to be compared to benefits and risks experienced several decades in the future?), estimating the probabilities associated with uncertain events so that expected values of their benefits and costs can be calculated, and relating each component of a program to the potential benefits. Nowhere in the discussion of the benefits of recombinant DNA research is there discussion of how current and proposed research projects will contribute to capturing these benefits, and whether the guidelines and the NIH research budget set up the proper incentives for molecular biologists to pursue the lines of research that will make the greatest contributions to achieving these objectives. Nor is any discussion to be found on the relationship of alternative safety standards, including those set by the guidelines, to the cost of acquiring the knowledge that is needed to commercialize DNA recombinations. Nor is there an analysis of how alternative safety standards affect the kinds of benefits that ought to be pursued most vigorously and, consequently, the particular lines of basic research that ought to be emphasized.

Another essential element to a benefit-risk analysis is to explore the alternative uses of the same resources and the alternative means to satisfying the same ends. Presumably a ban on recombinant DNA research would cause molecular biologists who do this work to switch to other kinds of genetic research. While the gross cost of this switch would be the knowledge that can only be attained through recombinant DNA research, the net cost would be less since, presumably, other lines of genetic research would progress more rapidly. A question that requires answering in a benefit-risk analysis is what benefits from other lines of research by molecular biologists are being sacrificed or delayed by devoting significant resources to recombinant DNA research.

Of course, the potential benefits of recombinant DNA research may also be reachable by other means. A precise statement of the benefits that might accrue from recombinant DNA research is that it may contribute to disease treatment, food production, and several other objectives, just as other lines of research may also make contributions in the same areas. A valid benefit-risk analysis would estimate the extent to which some expenditures on recombinant DNA research would increase the chance that society will capture these benefits for a given total expenditure on all paths to the same ends. For example, is a better way to reduce the death rate from cancer to seek cures for viral cancer through recombinant DNA research, or to expand research on environmental causes of cancer? Or, if in the long run insulin supplies are likely to run short, how should emphasis be divided among recombinant DNA studies, research on other synthetic processes, or expansion of supplies from animals?

Related to the question of the selection of a comprehensive research strategy for achieving the objectives mentioned in justifying recombinant DNA research is the question of the best timing for various activities that might contribute to the attainment of these ends. In particular, one alternative to an immediate, up or down decision on DNA recombination is

272

to delay all or part of it.  The discussion about the costs and benefits of further delays in pursuing this research has focused on the costs-- postponing for the period of the delay the date at which the benefits will be reaped and losing national prestige if scientists in other countries produce successful research before Americans do.  But the delay in bene- fits is trivial, indeed, if they are in any case unlikely to accrue for decades.  On the other side, delay can be especially valuable if an activity has some chance of causing a catastrophic, irreversible event and if further investigation of methods to reduce the chance and impact of the event is likely to pay off in a relatively short period of time. At least two issues in the debate about recombinant DNA suggest that these conditions do apply in this case.  One is the possibility that research that is as informative as the research now under way could be performed in a few years if attention were focused on developing safer sources, vectors, and hosts.  The other is the disagreement among molecular biol- ogists as to whether there is a natural barrier to DNA recombinations between prokaryotes and eukaryotes.

In any situation involving risks that have unknown dimensions, one potential benefit of a research project is to acquire more information about risks without actually having to be exposed to all of them.  Be- cause technical experts disagree about the potential hazards of recombinant DNA, one criterion for evaluating current research ought to be the extent to which its results will contribute to society's ability to comprehend and minimize the risks of further research.  An unfortunate feature of benefit-risk analysis is that its practitioners tend to think in terms of adopting an optimal long-term solution to the problem of decision making under risk, as if guns were being held to the heads of decision makers to make final decisions on the basis of current information.  But if some research activities are known to avoid risks that are endemic to other activities but, at the same time, to contribute to the informa- tion upon which further risk assessments will be made, it may make sense to pursue the former activities even if their direct contribution to ultimate societal objectives is less than that which the latter activities are likely to make.

Is Benefit-Risk Analysis Appropriate?

Perhaps more fundamental than the preceding issues concerning the require- ments of a valid benefit-risk analysis is whether DNA recombinant research ought to be evaluated in this way at all.  To approach the problem with this frame of reference is to adopt the view that research is primarily an investment to achieve normal economic ends.  If so, the first immediate question is whether government should be involved at all in commercial- izing molecular biology.  If, as seems likely, hybrids created from recombinant DNA research are patentable, is it plausible that drug companies and other private firms lack sufficient incentives to develop hybrids and, therefore, that government must be the principal source of support for this research?  Moreover, if these companies do lack sufficient incentives, is it not more appropriate that government should subsidize corporate research in this area on the grounds that private industry is more likely to pursue cost-minimizing programs that are more closely directed to achieving commercialization than is the scholarly research community?

The principal consequence of selecting research projects on the basis of their returns as investments is that the basis for selecting them must be other than the scientific interests and curiosities of the researcher. A project can be of scientific interest because it requires a display of virtuoso technique, because it resolves a technical dispute that has no practical consequence, or because, after the fact, it turns out to have provided some new insight that was completely unanticipated. In these respects, research is more like a novel or a work of art than like a capital investment, and these features are likely to be ignored if research is to be regarded as another form of investment.

Society may commit public funds for research for numerous reasons: it may value more knowledge for its own sake, it may regard research as a necessary cost of maintaining a system of higher education (without the possibilities for research, could as many good medical schools be operated?), or it may be governed by a winning political coalition that includes the research community, along with the beneficiaries of tax shelters, and that succeeds in redistributing income in favor of itself. Whatever the reason, the resulting structure of research will be quite different if projects are not selected strictly on the basis of their *ex ante* likely contribution to some particular instrumental end. In particular, the system of diverse, independent research scholars who individually control their selection of research topics and collectively determine how research dollars shall be spent, which projects are most interesting, and what proposals should be publicly supported, is not consistent with an instrumental, investment-oriented, "Big Science" model of research. Moreover, the relationship between society and research is far different in the two systems. In the science-as-art model, society may retrospectively alter financial support for research on the basis of several performance indicators--the state of the system of higher education, the amount of interesting new scientific information being reported, and the spin-offs of basic research for practical ends--but the main issue with regard to the selection of future research projects is whether they conflict seriously with other activities that contribute to society's welfare. This model is very close to the model of personal behavior in a free society; scientists are free to pursue whatever lines of inquiry they find interesting as long as they avoid direct harm to others.

Recombinant DNA research takes on a different light when viewed against the science-as-art template. First, a particular activity that constitutes a relatively tiny part of research in general and that is a source of anxiety for large numbers of people, for whatever reason, is likely to lose public financing. Second, if the risks associated with a particular line of research are real, but nevertheless offset by potential benefits, the mechanism of undertaking the research is likely to be quite different than the customary academic research mode. In particular, government will seek ways to do as little of the risky research as possible while capturing maximal instrumental benefits, to control research methods very closely, and to become more deeply involved in making *ex ante* judgments about the instrumental value of research proposals. Third, regardless of one's feelings about the ethical aspects of assigning burdens of proof, if the instrumental benefit of a line of research is not regarded by nonexperts as worth the risks that they perceive, the scientific community will bear the burden of proof in reducing uncertainties about the extent of the risks involved.

CONCLUSIONS:  THE FUTURE OF DNA RECOMBINANT RESEARCH

Public policymaking on recombinant DNA will be influenced by many factors other than the ones discussed above. Certainly if Congress perceives a significant risk in recombinant DNA, it will move to adopt more stringent controls than are likely to arise from a self-regulatory process administered by an agency with no enforcement authority. But other realities will also influence the outcome. Perhaps the most important is that not all of the research--or even most of it--is likely to take place in American nonprofit research institutions. This means that regulations based upon the role of the federal government as the principal source of research funds for the nonprofit sector may deal with only the tip of the iceberg, particularly on a global scale.

Institutions are already in place that deal with the kinds of hazards associated with recombinant DNA research. For example, OSHA and the Environmental Protection Agency (EPA) can enter the arena without further legislative mandate, and almost surely will if they perceive recombinant DNA research to be risky and, in particular, if private industry begins to pursue this research with vigor. Moreover, if private industry is subjected to significant regulation in this area, universities will not be far behind. It did not take long for OSHA and the Equal Employment Opportunities Commission to include universities within the ambit of the regulatory policies that they pursue.

The first choice facing the government is whether to support recombinant DNA research. Although public officials may believe that society would be better off if the research did not take place, they really cannot accomplish this objective on a worldwide scale. Consequently, the decision about financing must be partly strategic: How can the federal government provide financial support in such a way that the resulting research is least threatening? Several considerations come to mind in this regard. One, as discussed in this paper, is to support research on the development of containment systems and source-host-vector combinations that are safer than those that are currently available. Another is to be far more generous in supporting the less risky lines of recombinant DNA and other research in molecular biology research than in supporting more risky projects. Still another is to attract as much of the research and commercial development into the public sector as possible by being perhaps unnecessarily lavish in providing funds to create the optimal research environment for essentially any legitimate molecular biologist.  This would maximize the extent to which knowledge about molecular biology is in the public domain, and therefore minimize private incentives to do the work by reducing the likelihood that private research would produce proprietary information.  It would also give the government greater control in directing the lines of genetic research that are pursued.

A second area of decision making involves the selection of a system of controls on recombinant DNA research.  An immediate step is to establish regulations regarding commercial uses of recombinant DNA before the first commercial use emerges.  The nature of these regulations will affect the incentives private industry has to pursue this research; obviously a ban on commercialization backed up by criminal penalties represents an extreme action that would immediately stop most private research in the field.  Alternatives include licensing and inspection procedures through an agency such as FDA, EPA, or OSHA.  Serious examination of the problems of preventing severe accidents with commercial quantities of recombinant DNA hybrids will contribute to more than the development of a regulatory policy that is probably inevitable.  It will also shed additional light on the nature of the risks of this research in general and upon the likelihood that extensive commercialization is a real possibility.

With regard to scholarly research, Robert Sinsheimer's proposal to limit federally supported research to government-owned facilities deserves serious consideration.  First, a few large government facilities are much easier to control than a diffuse system of small laboratories with differing designs and procedures.  Second, such a system relieves universities of bearing most of the risks of the actions of their molecular biologists.  Third, in government facilities it will be easier to develop a coherent system of monitoring the performance of containment systems for the purpose of reevaluating risks and altering standards and procedures.  Fourth, because government regulation must be accompanied by complex administrative procedures to satisfy constitutional protections of due process, regulatory rules and standards are difficult to change.  Government laboratories need not be subjected to these formalities and, consequently, can change safety procedures quickly in response to new information and contingencies.

In order to make rational decisions on recombinant DNA, policymakers will need expert analysis and advice.  To avoid some of the problems of inadvertent delegation of control to the experts, policymakers should consider assembling a panel of near-experts whose training will enable them to comprehend the technical issues but whose professional pursuits do not involve recombinant DNA methods, and who thus suggest by self-selection that they are, on average, more representative in their tastes and risk assessments regarding recombinant DNA research than are the experts.  The job of this group would be to translate and evaluate the technical case of the experts, and to raise further questions that may have been overlooked in the debate.

In sum, federal policy should be based upon the notion that a well-designed program can redirect the focus of research in ways that reduce societal risks.  At the same time, the federal government should probably abandon, at least for the present, establishing policy towards recombinant DNA on the basis of future commercial spin-offs.  Instead, for a while the focus should remain on guaranteeing that as much of the research as possible will take place in carefully controlled environments and will contribute both to advancing basic knowledge about genetics and to reducing the uncertainties and risks surrounding research in this area.

BIBLIOGRAPHY

Bernardi, G.  1976.  Expression of eukaryotic genes in prokaryotes. Nature 259:173-74.
Chang, A. C. Y., and S. N. Cohen.  1974.  Genome construction between bacterial species *in vitro*:  replication and expression of *Staphylococcus* plasmid genes in *Escherichia coli*.  Proc. Natl. Acad. Sci. U.S.A. 71:1030-34.
Committee on Recombinant DNA Molecules (National Research Council).  1974. Potential biohazards of recombinant DNA molecules.  Proc. Natl. Acad. Sci. U.S.A. 71:2593-94.
Enquist, L., D. Tiemeier, P. Seder, R. Weisberg, and N. Sternberg.  1976. Safer derivatives of bacteriophage $\lambda$gt·$\lambda$c for use in cloning of recombinant DNA molecules.  Nature 259:596-98.
Genetic engineering focus of joint R & D.  Chem. Eng. News, October 21, 1974, p. 36.
Industry wary about genetic guidelines.  Chem. Eng. News, June 7, 1976, p. 7.
Lawrence, E.  1974.  Science and politics of molecular biology.  Nature 251:94.
Lewin, R.  The future of genetic engineering.  New Sci., October 17, 1974, p. 166.
McBride, S. D., and B. Knickerbocker.  Research limits unprecedented. Christian Science Monitor, February 9, 1977, p. 3.
Murray, K.  1974.  Alternative experiments.  Nature 250:279.
Murray, N. E. and K.  1974.  Manipulation of restriction targets in phage $\lambda$ to form receptor chromosomes for DNA fragments.  Nature 251: 476-81.
Norman, C.  1975.  Genetic manipulation recommendations drafted.  Nature 258:561-64.
Norman, C.  1976.  The public's case is put.  Nature.  259:521.
Playing with genes.  The Economist, November 8, 1975, p. 18.
Rawls, R. L.  NIH genetic guidelines get mixed reviews.  Chem. Eng. News, July 5, 1976, pp. 29-30.
Sherratt, D. J.  1974.  Bacterial plasmids.  Cell 3:189-95.
Sherratt, D. J.  1976.  Biological safeguards in genetic engineering. Nature 259:526-27.
U.S. Department of Health, Education, and Welfare.  1976.  Recombinant DNA research guidelines.  Fed. Regist. 41(131):27902-43.

U.S. House of Representatives. 1974. Genetic engineering: evolution
    of a technological issue. Report for Subcommittee on Science, Re-
    search and Development. Government Printing Office, Washington,
    D.C.
Wade, N. 1975. Recombinant DNA: NIH group stirs storm by drafting
    laxer rules. Science 190:767-69.
Wade, N. 1976. Recombinant DNA: guidelines debated at public hearing.
    Science 191:834-36.
Wade, N. 1976. Recombinant DNA: the last look before the leap. Science
    192:236-38.
Wade, N. 1977. Dicing with nature: three narrow escapes. Science
    195:378.

# DISCUSSION

KURT MISLOW, Princeton University: The duty of this Panel for Inquiry
is presumably to focus the discussion and to look for consensus. It
would seem fairly difficult to do that in light of what has been referred to as a polarization of opinions, and which I regard more as
a dialogue of the deaf, because different ground rules are employed
in this discussion that we have heard.

But there are in fact two areas of consensus that can be dis-
cerned. One of them is that the only certain benefit is knowledge,
and I quote from Maxine Singer's *Science* paper that was cited by one
of the speakers, and the other being that there is not enough information to quantify reliably the risks and benefits. I hope we can
all agree with that at least.

Well, just to show you what sort of contrary fellows we are at
Princeton, I would like to dissect the word "benefit" in the statement "the only certain benefit is knowledge." If I had been asked
why I am in science it would be because it is a lot of fun, because
it is a life's work, because I can't think of anything more exciting
to do. And I think I speak for all of us in this room who are scientists that we are in this game because it is just fun, and we would
rather do this than anything else. To claim that there is a further
benefit is to assume that we understand the consequences of what we
have learned, and I claim that the value judgments in this respect
must necessarily be indefinitely based, at best. I claim that knowledge is, in a sense, indeterminant, since it can be, as has been
pointed out, applied in many, many ways.

I will undoubtedly provoke cries of inquisition and the like, but
I must nevertheless force myself to say, and I only say so because
I have tenure, that I don't agree that freedom of inquiry should be
limited only if actual hazards are perceived. I do not agree that
increased human knowledge is of paramount importance. I do not agree

that the real enemy is ignorance. I think these are trademark shibboleths which everybody accepts without questioning. I can think of lots of examples where knowledge is extremely dangerous. And in the search for knowledge, you have to ask what you are going to do with the knowledge once you have acquired it.

A recent and notorious example of this is the so-called research into the genetics of IQ. Even if we leave aside the question of fabrication of data, which has recently been exposed, there still remains the question if such research is done and after it has been done, what is going to be done with the results? I would ask all of you to think about that rather seriously. It was pointed out by one of the workshop chairmen, Professor Green, that other persons do not regard scientific research as necessarily being an unmitigated good. But no one present has articulated this view. Well, I am here to articulate it. In other words, when I use the word benefit, I also ask "Whom does it benefit?"

The next point I would like to raise is that although I regret polarization, I do not regret politicization. I think this is basically a social, political, and economic problem. It is not in the hands of academia; it is not in the hands of industry. It is in the hands, in fact, of the people. We the people, the lay people --and I consider myself by definition a lay person because I am not a molecular biologist--must depend on experts as we do, for example, when we train the military. Nobody questions civilian control of the military. Why do people question civilian control or lay control of what is a powerful technique that might alter our very society, as has been pointed out by many people?

It has been said that there has been too much wringing of hands. Well, I might say that the nuclear physicists have been wringing their hands for many years since Hiroshima. I prefer to wring my hands before the fact rather than after, and I think this is something that we might consider as well.

Finally, I have to say one other thing. I feel that much has been made of the somewhat irrational, if you like, and emotional nature of the opposition. I am referring to the broad sense of opposition that one feels on the part of many people here for all kinds of different reasons. The opposition is, in fact, necessarily emotional, necessarily irrational because it is not by the rules of the game that these objections are being made. We are not providing a data base, drawing conclusions, and providing inferences and so forth. We are operating from a gut feeling. That is to say we, I mean many of us, operate from a gut feeling, which is nevertheless very real. To ignore the reality of that gut feeling, I think, is to be, in the last analysis, unscientific. It is there. Perhaps--and I am being a little bit romantic now--but maybe our genes are trying to tell us something. This is something that goes back to our childhood. It goes back to the fairy tales that we heard when we were little kids. It is something very fundamental, and I think we are making a tremendous mistake if we ignore it.

JUNE GOODFIELD, Visiting Professor, Rockefeller University [Ms. Goodfield
has added an analysis of this Forum to her new book entitled *Playing
God: Genetic Engineering and the Manipulation of Life*, to be published by Random House in the fall of 1977]:  Looking back over the
three days of this Forum and the two years of research that preceded
the actual writing of a book on this subject, I am reminded of an
episode in England's civil war when the decision was being taken by
Cromwell and the members of Parliament whether they should or should
not chop off King Charles' head.  Now, this may seem to you in the
days of the republic to be a matter of very small moment.  Actually,
because of the divine right of kings, which in many ways seems to
parallel the divine right to search for truth, it was an exceedingly
important episode, and an exceedingly important decision that had to
be made.

Naturally, not everybody agreed that Parliament had either the
right or the confidence, let alone the hubris to do this, and there
was a minority report.  And the man who put in the minority report
said:  "I pray God you consider for a moment the possibility that you
may be mistaken."

Now, when this debate started from Asilomar we all expected that
everybody would mount their own favorite hobbyhorses very rapidly,
gallop off wildly, or not even wildly but rationally in the directions
of their own horses, firing cartridges as they went, some of which
have turned out to be blank ones.  What I think many of us did not
foresee and were very unhappy to see was the speed with which those
hobbyhorses metamorphosed into war horses and battleaxes started
to be wielded.  I think that the present hardening of the lines,
the present way in which people have tended now to come to feel that
this is an issue over which there are victories to be won or victories
to be lost, points to be gained and points only to be conceded is a
very, very dangerous one.  I hope that this can stop here and now,
because we are already lost if we continue in this way.

I think that what we do have to do is to try to regard this whole
issue not as a battlefield, but much more as an opportunity to reassess
the nature of the social contract between science and society.  This
issue I think under no circumstances is going to go away for that very
reason, and I do think that we could, if we wished to, use it in order
to return to a state which science once enjoyed, where Rousseau's
original social contract was in fact manifested--that is, a state of
mutual benefit between two parties.

ALAN C. KAY, NIH:  I work in the laboratory of Dr. Maxine Singer at NIH,
which does not mean to say that I hold the same views as she does on
the use of recombinant DNA research, and of the use of the guidelines.
In fact, I cannot say at the moment that the position I hold is for
complete cessation of the use of recombinant DNA technology, but it is
also true that my position is not very far from that and may vary in
the future.

This morning I raised two points about workshop 8, and I am going
to some extent raise them again.  The first point that I tried to

raise was that the industry attitude toward the NIH guidelines is more lip service than anything else.

The second point which I raised this morning was that in the report of Dr. Chakrabarty he expressed or stated some very detailed types of guidelines which could be applied to industry and also some types of enforcement that could be applied to these guidelines, and that in fact such a discussion had not been held and that therefore there could not possibly be any consensus about that.

I don't know how many people here could imagine that in the future, in six months, in a couple of years, in several years, when there are legislative hearings either at the federal, state, or local level, some industrial representative will not appear before such a committee with a photocopy of Dr. Chakrabarty's report, and saying here is the consensus of the Forum of the National Academy of Sciences. So I think that Dr. Chakrabarty should make it clear that there was in fact no consensus within this workshop.

LEON JACOBS, Associate Director for Collaborative Research, NIH: You have just seen that in the intramural program at NIH somebody can be outspoken against the guidelines and still retain his position.

I want to return to Professor Wald's allegations of yesterday that study sections are influenced by opposition to the guidelines. I went to the Division of Research Grants this morning and asked them to identify grants awarded to three outspoken opponents of DNA research in the Boston area, and I just got the information at a little after 3:00 this afternoon. Those three individuals have seven grants, all but one of which was awarded after the time when they became outspoken. So I think that these individuals ought to allay Dr. Wald's fears on that subject.

A. M. CHAKRABARTY, Microbiologist, General Electric R & D Center: I would like to respond to what Dr. Kay said in terms of my unfortunate use of the word *consensus*. I have a copy of the report that I gave to Dr. Hamburg in the morning. It was given to him before I had the opportunity to speak. If you go through the report you would find that nowhere have I used the word that we have come to consensus. It was on a spontaneous basis I said that. What I really meant to say is that in an attempt to come to consensus the following salient features of the discussion came through. I apologize if I am misunderstood that I said we came to consensus. We didn't come to consensus. As a matter of fact it is very difficult to come to consensus when you have so many diverse opinions.

There is no way I can determine whether the industry is giving lip service to what they are saying. I would imagine that would be up to the Congress to find that out.

FREEMAN J. DYSON, Institute for Advanced Study, Princeton, a member of the Panel for Inquiry, asked that the following commentary be entered into the record at this point:

## JOHN MILTON AND EXPERIMENTAL MICROBIOLOGY

John Milton wrote a speech 333 years ago defending the freedom of the press, which also speaks to our present concerns. I have collected some of the salient passages from the speech, in the hope that this may help us to take a longer view of the problems we are discussing. I am suggesting that there is an analogy between the seventeenth-century fear of moral contagion by impious and soul-corrupting books, and the twentieth-century fear of physical contagion by pathogenic microbes. In both cases, the fear was neither groundless nor unreasonable. How far the analogy between seventeenth-century books and twentieth-century experiments is valid, I leave to you to judge for yourselves. For the human and institutional aspects of our problem, at least, the analogy seems to me to be real.

For this is not the liberty which we can hope, that no grievance ever should arise in the Commonwealth--that let no man in this world expect; but when complaints are freely heard, deeply considered, and speedily reformed, then is the utmost bound of civil liberty attained, that wise men look for.

[This is the seventeenth-century version of multiple bureaucratic regulation.] Sometimes five Imprimaturs are seen together, dialogue-wise, in the piazza of one title-page, complimenting and ducking each to other with their shaven reverences, whether the author, who stands by in perplexity at the foot of his epistle, shall to the press or to the sponge.

[Milton was willing to suppress books that were openly blasphemous, just as we are willing to ban experiments that are demonstrably dangerous. The difficult problems arise with "things uncertainly and yet equally working to good and to evil."] Suppose we could expel sin by this means; look how much we thus expel of sin, so much we expel of virtue; for the matter of them both is the same; remove that, and ye remove them both alike. This justifies the high providence of God, who, though he commands us temperance, justice, continence, yet pours out before us, even to a profuseness, all desirable things, and gives us minds that can wander beyond all limit and satiety. Why should we then affect a rigor contrary to the manner of God and of nature, by abridging or scanting those means, which books freely permitted are, both to the trial of virtue, and the exercise of truth?

It would be better done, to learn that the law must needs be frivolous, which goes to restrain things, uncertainly and yet equally working to good and to evil. And were I the chooser, a dram of well doing should be preferred before many times as much the forcible hindrance of evil doing. For God sure esteems the growth and completing of one virtuous person, more than the restraint of ten vicious.

[Here Milton describes in vivid language the difficulty of finding competent and conscientious people to staff a regulatory agency.] Another reason, whereby to make it plain that this Order will miss the end it seeks, consider by the quality which ought to be in every licenser. It cannot be denied but that he who is made judge to sit upon the birth or death of books, whether they may be wafted into this world or not, had need to be a man above the common measure, both studious, learned, and judicious; there may be else no mean mistakes in the censure of what is passable or not; which is also no mean injury. If he be of such worth as behoves him, there cannot be a more tedious and unpleasing journey-work, a greater loss of time levied upon his head, than to be made the perpetual reader of unchosen books and pamphlets, ofttimes huge volumes. There is no book that is acceptable unless at certain seasons; but to be enjoined the reading of that at all times, and in a hand scarce legible, whereof three pages would not down at any time in the fairest print, is an imposition which I cannot believe how he that values time and his own studies, or is but of a sensible nostril, should be able to endure.

In this one thing I crave leave of the present licensers to be pardoned for so thinking; who doubtless took this office up, looking on it through their obedience to the Parliament, whose command perhaps made all things seem easy and unlaborious to them; but that this short trial hath wearied them out already, their own expressions, and excuses to them who make so many journeys to solicit their license, are testimony enough. Seeing, therefore, those who now possess the employment, by all evident signs wish themselves well rid of it, and that no man of worth, none that is not a plain unthrift of his own hours is ever likely to succeed them, except he mean to put himself to the salary of a press corrector, we may easily foresee what kind of licensers we are to expect hereafter, either ignorant, imperious, and remiss, or basely pecuniary. This is what I had to show, wherein this Order cannot conduce to that end, whereof it bears the intention.

283

[The connection between the silencing of Galileo and the general decline of intellectual life in seventeenth-century Italy has been made much of by modern liberal historians. Here we see that the connection was not invented by the historians but was also obvious to a contemporary eye-witness.]  And lest some should persuade ye, Lords and Commons, that these arguments of learned men's discouragement at this your Order are mere flourishes, and not real, I could recount what I have seen and heard in other countries, where this kind of inquisition tyrannises; when I have sat among their learned men, for that honor I had, and been counted happy to be born in such a place of philosophic freedom, as they supposed England was, while themselves did nothing but bemoan the servile condition into which learning amongst them was brought; that this was it which had damped the glory of Italian wits; that nothing had been there written now these many years but flattery and fustian.  There it was that I found and visited the famous Galileo, grown old, a prisoner to the Inquisition, for thinking in astronomy otherwise than the Franciscan and Dominican licensers thought.  And though I knew that England then was groaning loudest under the prelatical yoke, nevertheless I took it as a pledge of future happiness, that other nations were so persuaded of her liberty.

[As shown in the following passages, Milton's patriotic pride in the intellectual vitality of seventeenth-century England is a sentiment that twentieth-century Americans can share.]  Lords and Commoners of England, consider what nation it is whereof ye are, and whereof ye are the governors; a nation not slow and dull, but of a quick, ingenious, and piercing spirit, acute to invent, subtle and sinewy to discourse, not beneath the reach of any point the highest that human capacity can soar to.

Nor is it for nothing that the grave and frugal Transylvanian sends out yearly from as far as the mountainous borders of Russia, and beyond the Hercynian wilderness, not their youth, but their staid men, to learn our language and our theologic arts.

What should ye do then, should ye suppress all this flowery crop of knowledge and new light sprung up and yet springing daily in this city; should ye set an oligarchy of twenty engrossers over it, to bring a famine upon our minds again, when we shall know nothing but what is measured to us by their bushel?

For when God shakes a kingdom with strong and healthful commotions to a general reforming, 'tis not untrue that many sectaries and false teachers are then busiest in seducing; but yet more true it is, that God then raises to his own work men of rare abilities, and more than common industry, not only to look back, and revise what hath been taught heretofore, but to gain further, and go on some new enlightened steps in the discovery of truth.

## DAY I AND DAY II

Tracy M. Sonneborn

*Distinguished Professor Emeritus of Zoology, Indiana University*

This summarizer will deal only with the first half, perhaps the less controversial half of the Forum. However, I am going to talk about myself a little bit because I think it is necessary to know, if you can, what kind of person it is who comes to the conclusions and says the things that I propose to say.

Like my fellow scientists I was and remain full of what might be called religious wonder and awe at the universe and all that is in it, including human beings and all living organisms. And I had and still have an abiding, if naive, faith that men and women can bit by bit go on indefinitely decreasing the immensity of our ignorance by scientific investigation and by humanistic insights.

It has not been easy to maintain that faith, or to recover it after lapses, in view of the bad times that Chargaff and Wald and I and others of our generation have lived through--the Great Depression, the Hitler years, the Lysenko period, the McCarthy period, the A-bomb, Vietnam, Watergate, and now anti-intellectualism. Some experiences of the last few days haven't made it any easier.

When asked to summarize this half of the Forum, foolishly as it may seem, I could not say no. Why? Not because I have the saintly qualities of Dr. Hamburg, as demonstrated on the first evening, but because of my past experiences. I began teaching genetics forty-four years ago, and

year by year have tried to assimilate and to excite students about its
steady stream of marvelous advances. As Dr. Hotchkiss told you earlier
today, I was sensitized to the societal implications fourteen years ago.
Then several years ago I started at Indiana University a course on human
and general genetics to which was admitted only those college students
who had no college biology. So I share with George Wald and many others
the conviction that scientists can convey to nonscientists, and to those
who consider science incomprehensible, the major scientific facts and
principles and the spirit and the faith of science. I accepted my assignment here because I thought I would be going through a comparable exercise.

Well, it is impossible to summarize in twenty minutes the large variety
of facts and errors, of fears and misunderstandings, of guesses and calculations of probabilities, of arguments and counterarguments, of protests
and diversions, of satisfactions and dissatisfactions, of proposals and
counterproposals. Here instead I shall report the impact of this Forum
on this one observer.

First, it seems to me that this Forum has brought out very clearly the
depth and range of the issues raised in various people's minds and hearts
by recombinant DNA technology, of the ways it can be used, and by the
variety of "lobbies," to borrow Kendrew's term, that have a passionate
interest in them and in the broader issues to which they lead. Unfortunately the Forum has also shown that it is quite impossible, in ten-
or twenty-minute presentations and short comment and question periods,
to give adequate opportunity for full expression, much less to follow
through in the more arduous and, alas, perhaps impossible effort to resolve differences.

To accomplish such objectives it seems to me there are only two possible alternatives. Either an open-ended, long series of sessions, or
recognizing human limitations of endurance, strictly limiting each of a
series of forums to one or a few specific issues, and being absolutely
tough in keeping the discussion to the issue at hand.

Now, that doesn't mean that I fail to see that the issues are inter-
related, or that any particular ones are irrelevant, but by jumping
freely from one to another, we really fail to do justice to any.

Second, this Forum and much that preceded and led up to it is a
recognition that the public is vitally concerned with some areas of
scientific research, that it should be appropriately informed about it
in time to fulfill effectively its responsibility in arriving, in discourse with the scientists, at judgments and actions about choices and
regulations. I, for one, and many agree with me, do not doubt that most
of the molecular biologists tried to do this with the best of motives as
Maxine Singer and others clearly recounted, and in the only ways they
then saw available to them. They had no real precedents, no experience
to go on. With hindsight we may think we see mistakes and oversights,
but as Callahan pointed out, there was no clearly visible apparatus to
do the job better given that something had to be done at once. And
they recognize that the result, the guidelines, are tentative and subject
to change.

Moreover, and this has been hinted by others, a full recognition of
what the public is is not yet clearly recognized; but we surely have had

some eye-openers here. The public, we are learning, is a great composite of varied and often conflicting, passionately conflicting, interest groups, scientists among them. No matter what one's prejudices or biases may be about this or that interest group, and I don't pretend to be free from such biases, all those that think they have an interest have to be given full opportunity for input. While there seems to be no possibility of reaching a consensus on this or on any other complex issue, if we take democracy seriously, base has to be touched with all of the interest groups, or they may well become implacable enemies. True, they may anyway, but they certainly will if kept out. The Forum met that challenge with remarkable patience and openness.

Third, one thing depressed me more than any other in this Forum because I am closer to it. I am still naive and idealistic enough to hope that scientists, by their training and experience, have respect for facts, are accustomed to grasp probabilities, and are willing and able really to listen to each other, especially to others who have a different expertise, and even occasionally to change their minds in view of the evidence and reason. This did not always happen here.

Scientists came with their minds made up, not yielding on anything of importance. I believe they all have their conception of the public good passionately at heart, but sometimes they seem to this observer to have their minds, even their ears, tightly closed, and to lean on sometimes demonstrably false facts, or remain deaf to cogent reasons. I do not hold that scientific questions are settled by majority vote. Nor do I hold that the minority either is always wrong or is always right and wise. Nor am I impressed by the scientific distinctions of the disputants. There are equally distinguished scientists on both or all sides. As Gilbert and Sullivan put it, where everybody is somebody, nobody is anybody.

It is no good, and unworthy of scientists, to impugn motives or run down opponents personally. The purely scientific issues can be scientifically settled only by facts; or, when facts are lacking, by getting them if possible; and if not possible, by the rule of reason. Until scientists discussing the present scientific issues do that, they can only confuse the issues, confuse and frighten the public, and feed the burning fires of anti-intellectualism.

It is my impression and my faith that nonscientists can at least sometimes see through the obliquity of us scientists and come to a workable consensus, as it seems to me the lay committee at Cambridge did.

Now, I want to summarize briefly the impressions made on me concerning some substantive issues of the Forum. First, however, I must disclose the extent of my bias on these before I came, for none of us can be completely free from bias. My whole research career, as those who know it will agree, has been one of challenge to generally accepted dogmas. I have never long been in the mainstream of genetic research, and I am certainly far from a molecular biologist. You would not then be surprised if this gives me an initial bias toward minority opinions. Nevertheless, I came to the Forum and to my preparations for it about as uncommitted, I think, as someone could possibly be. When I read and/or heard the pros, they seemed convincing, until I read and heard the cons. And then the cons swayed me. But I remained throughout intensely excited by the sheer intellectual elegance of the new recombinant gene technology in spite of my being an old-fashioned geneticist, and by its eloquently stated fundamental uses and possible uses in attacking old and basic problems of life: the evolutionary development of the organization of genes and chromosomes, the mechanisms that control their actions, and how these mechanisms function in the orderly processes of individual development. The possibilities of getting at these problems excite me. The many suggested possible practical applications did not equally impress me. They may or they may not come to pass. They may impress cost-benefit analyzers, but they do not carry that much weight in my mind. I suspect that the most important actual future applications may not even be now foreseeable or imaginable.

But again, I have faith that sooner or later there will be important applications, but best of all, far best of all, much insight into nature and human beings that society in the long run will bless, not curse us for.

Now, there is no *a priori* necessity that I can see to use these methods for human genetic engineering. Scientists and society can prohibit that if they want when the time comes, when it should, if ever, be possible. Though I am a biologist I am, like many of you, also a humanist, caring greatly for human good and protection. I therefore agree with the critics and also many of the proponents that high priority should be given to protection of the public as well as to the laboratory workers by research directed toward finding or developing ever-safer organisms for culturing the segments of DNA, as is being done by Curtiss; that training courses should be developed for lab workers; that we should lean over backwards in safety measures to assure minimal, reasonable hazard.

I was impressed in the accounts that were given here, particularly by the long experience of epidemiologists and others in the medical field. And being a geneticist, I was much more impressed than the audience and the nongeneticists seemed to be by the genetic arguments put forth by Ayala and Davis. They seem to minimize the dangers of bacteria to which some DNA of other organisms, usually less than one-tenth of one percent of the bacterium's own DNA, is added; and to show that comparable although not identical additions occur naturally by unequal crossing-over, mutation, and allopolyploidy.

I cannot, therefore, share the fears of Chargaff, Wald, and others when they raise what they call the basic question of whether recombinant DNA research should be permitted at all. I was strongly persuaded by Cohen's distinction between the technology of DNA recombination per se and the uses that are made of it. It seems to me absurd to make a blanket restriction on all uses when some of them do no more than happens naturally, as among bacteria that are recombining their genes in nature.

Moreover, as Tooze has clearly pointed out elsewhere, complete prohibition is unenforceable unless one is willing to accept a totalitarian state, a price few would be willing to pay. Moreover, Tooze, Callahan, and others clearly made the point that effective control in the last analysis depends on the allegiance of scientists, and that is unobtainable if restraints, even legally binding restraints, are excessively unreasonable.

In the end the public, whatever that is, has to decide, as Callahan said, whether it will opt for slowness, caution, and repression by fear of all nonzero risks, or whether it is willing to go along with Lowrance's analysis defining safe as acceptable risk, and opt for scientific nerve and boldness. The basic public issue is whether to support what I consider the high adventure of science for its intellectual and cultural values, and possible but unsure practical applications, or whether to play it absolutely risk-free. In other words, is society willing to pay the necessary price?

Now, the pessimists warn that anything that can happen will happen, no matter how small the chance of its occurrence, and they set forth a great catalogue of imagined catastrophes, modeled on the fiction of the Andromeda strain. The optimists, willingly banned from the most hazardous experiments, hold that other hazards are safely excluded by acceptable restraints. And I am quite sure that both the optimists and the pessimists think that they are the real realists.

Now I happen to think that the public are the realists, and I still have confidence, though sometimes it falters, in their long-run, collective judgment. They will, in my opinion, not run the chance of throwing out this fine new baby with the bathwater. They will want reasonable safeguards, and they will want this new route for scientific progress to go forward as part of the high adventure of science in keeping with the words attributed to Prometheus, in referring to the arts and sciences, as inscribed on the panel outside of this room in the Great Hall: "I made them, the people, to have sense and to be endowed with reason."

## DAY III

### Donald Kennedy

*Commissioner, Food and Drug Administration; Former Benjamin Scott Crocker Professor of Human Biology and Chairman, Program in Human Biology, Stanford University*

The risks and benefits of agriculturally related work using recombinant DNA techniques are not difficult to summarize. We cannot judge the magnitude of either, just as we cannot with the biomedically related work; and so the situation resolves itself into a contest of best (or worst) cases. That evaluation is not intended to brush aside either presentation. The potential that resides in the chance that we may improve biological nitrogen fixation by incorporating enhanced genetic capacity into soil bacteria or (more remotely) higher plants is impressive, even though, as Dr. Valentine says, it may not be the likeliest path. Every recent analysis of the basic agricultural sciences has emphasized these opportunities, which loom larger as the industrial processes for nitrogen fixation become more energy- and capital-intensive. It is recognized, on the other hand, that the precariousness of agroecosystems may enhance the risk that misfires in this work could produce serious environmental mischief. Persuasive scenarios on both sides can be invented.

One point brought out in the debate deserves special attention, if only because it treats a recurring theme of the meeting. Agriculture and health, the great sister disciplines of applied biology, are both complex problems that cry out for social as well as for technical solutions. In the case of health, should we invest in delivery, health services research, and prevention, or should we invest in biomedical research? Such questions are often phrased as alternatives, as in the debate over the Report of the President's Panel on Biomedical Research. Similarly, in agriculture we have heard the hunger problem described as an income-distribution problem rather than a food production problem. As we attempt the marginal allocation of scarce resources, this either/ or view may be useful--a dollar for prevention rather than a dollar for cure. But the whole thing isn't margin, and I fear the argument has been taken farther than it will reach. Surely income distribution is a major factor in world hunger, given the edge of world per capita food production over world population. Every serious student of the problem I know concludes that aggregate annual food production in the developing countries must be augmented by 1.0 to 1.5 percent to keep pace with population projections and to make up present caloric deficits. Even highly optimistic assumptions about distribution improvements would require substantial gains; and, furthermore, most of these will have to be yield gains, because expansions in the cropland base are increasingly hard to come by. I commend to you the report of the Academy's World Food and Nutrition Study for a more detailed consideration of this question. Meanwhile, I think the fairest evaluation is this: the biological or technical gains are surely insufficient, but this hardly allows one to conclude that they are therefore unnecessary.

Now let me turn to the subject of regulation. You will understand that my own new responsibilities both reflect and intensify my concern with this matter. Indeed, one or two people have suggested that perhaps a kind of ecdysis will take place here, during which I will lay bare a philosophy of regulation that has been conveniently cloaked by the fact that no one had the slightest interest in it. I plan to take off one glove.

First, I observe that regulation is an arcane and possibly disturbing topic to many scientists, and to others as well. An indication of the limited way in which it is understood here, I would guess that the word regulation was used at least a hundred times in this room, before Roger Noll saved our collective sanity, with hardly a single mention of the terms *licensing, registration, liability, standardization, monitoring, detection, inspection, enforcement* or any other standard components of regulatory boilerplate. The guidelines are regulations in intent and practice, but only barely. As Dr. Davis emphasized, they have some special purposes, not least of which is to teach some microbiological technique to molecular biologists. They qualify as regulations because they contain sanctions and a compliance mechanism; but much of the latter depends upon voluntary action.

Why should there be more? The simple answer is because it is politically inevitable. I happen personally to believe, and it is just a guess, that this technology will turn out to be much freer of serious risk than it is held to be by the modal view in the scientific community. But I would not ask anyone to make public policy on that assessment, because there is simply too much uncertainty in it. At this point the controversy assumes the shape of the classic regulatory dilemma, which is just what makes it so interesting. It's got everything: uncertain but potentially significant benefits, unproven but potentially troublesome hazards, and unclear jurisdiction. The questions start to answer themselves. How much regulation are we going to have? Answer: As much as people insist on, in light of the kind of personal social value calculus that Roger Noll described for you.

Now I shall make a slight digression here to deal with a problem that troubles me, because it unnecessarily divides some of my friends. It is occasionally claimed that the scientists who called for the pause and first discussed the issue are responsible for the public concern. This line of argument goes on to draw a lesson from the incident: next time keep quiet, or you will stir unwanted controversy. It is a little like the mother's admonition: "Don't tell Junior *not* to stuff beans up his nose." I believe this view to be naive. It assumes a division between scientists and public that is now untenable. Even those who hold a low opinion of the capacity of lay persons to understand scientific issues surely can count the number of intelligent, scientifically trained persons who work in the media, on congressional staffs, or in organizations concerned about the public welfare. To suppose that risks associated with one of the most dramatic breakthroughs of the century, chronicled in the open literature with full attention from all these publics, would somehow escape political attention strikes me as a not fully worked-out position.

What is significant about the letter and the pause is that the scientists identified it *first*, and *acted*. To measure how much gratitude the rest of us owe them for that, you might want to consider what public attitudes we might now be encountering if, in addition to the present concern over the hazards, it could be legitimately claimed that scientists knew about them and kept quiet on purpose. Moreover, I think it is unfair to charge that the scientists are now trying to write the regulations. They couldn't, and they aren't.

A moment ago I said that there would be regulation, and tried to say why it will be more than the guidelines. Dr. Fredrickson chronicled the development of these under the Interagency Committee, on which I also served as OSTP representative. Regulations require public confidence in the processes by which they work. The voluntarism so central to the guidelines is not equally appropriate for all purposes; institutions differ, motivations differ, operations differ. What works for universities must work for industry; what works for R must work for D. Moreover, I think there are strong reasons for believing that a single set of federal regulations are much to be preferred over a mosaic of local ones. It is not just that epidemics are likely to be poor respecters of political boundaries, thus spreading risks as well as benefits broadly. It is also that consistency of practice, in most other cases we know about, assists the innovative process. Needed international coordination of procedures and practices will also be more easily handled with a federal regulation. In his treatment of the problems associated with monitoring, reporting, training of personnel, and standardization of practices, Dr. Dull provided other evidence for the desirability of national benchmarks.

In conclusion, I think a preference for this outcome over less attractive alternatives really should earn the last holdouts' support of the Interagency Committee's effort to agree on particulars of legislative initiative. I hope they will offer it fully, using their experience to suggest the most effective ways of keeping central track of what is happening, monitoring worker health, developing means of detecting escape, and emphasizing the significant differences in risk status between different kinds of recombinant DNA research. Even those who believe that the furor rests on misunderstanding, as it may, appear to agree that the alternative to good regulation is not *no* regulation, but *bad* regulation.

## GENERAL ADVISORY COMMITTEE

Robert McC. Adams, Professor, The Oriental Institute

Kenneth J. Arrow, James Bryant Conant University Professor, Project on Efficiency of Decision Making in Economic Systems, Harvard University

David Baltimore, American Cancer Society Professor of Microbiology, Center for Cancer Research, Massachusetts Institute of Technology

Arthur M. Bueche, Vice President, Corporate Research and Development, General Electric Company

Freeman J. Dyson, Professor of Physics, School of Natural Sciences, The Institute for Advanced Study

Donald S. Fredrickson, Director, National Institutes of Health

Gertrude S. Goldhaber, Senior Physicist, Brookhaven National Laboratory

Michael Kasha, Director, Institute of Molecular Biophysics, Florida State University

Daniel E. Koshland, Jr., Professor and Chairman, Department of Biochemistry, University of California, Berkeley, *Chairman*

Philip Morrison, Institute Professor, Department of Physics, Massachusetts Institute of Technology

John R. Pierce, Professor of Engineering, Steele Laboratory, California Institute of Technology

Alexander Rich, Sedgwick Professor of Biophysics, Massachusetts Institute of Technology

Frederick C. Robbins, Dean, School of Medicine, Case Western Reserve University

Lewis Thomas, President, Memorial Sloan-Kettering Cancer Center

Alvin M. Weinberg, Director, Institute for Energy Analysis, Oak Ridge

David A. Hamburg, President, Institute of Medicine, *ex officio*

Courtland D. Perkins, President, National Academy of Engineering, *ex officio*

## FORUM STAFF

| | |
|---|---|
| Robert R. White | *Director* |
| M. Virginia Davis | *Staff Associate* |
| Betsy S. Turvene | *Editor* |
| Marcie S. Lofgren | *Administrative Secretary* |

## PROGRAM COMMITTEE

Lawrence Altman, Medical Writer, *New York Times*

David Baltimore, American Cancer Society Professor of Microbiology, Center for Cancer Research, Massachusetts Institute of Technology

Paul Berg, Willson Professor of Biochemistry, Stanford University Medical Center

Daniel Callahan, Director, Institute of Society, Ethics and the Life Sciences

Freeman J. Dyson, Professor of Physics, School of Natural Sciences, The Institute for Advanced Study

Richard Goldstein, Assistant Professor, Department of Microbiology and Molecular Genetics, Harvard Medical School

Harold P. Green, Professor of Law, National Law Center, The George Washington University

Clifford Grobstein, Professor of Biology and Vice Chancellor for University Relations, University of California, San Diego

David A. Hamburg, President, Institute of Medicine, *Cochairman*

Howard H. Hiatt, Dean, Harvard School of Public Health

Stanley B. Jones, Staff Director, Subcommittee on Health, Committee on Labor and Public Welfare, U.S. Senate

Daniel E. Koshland, Jr., Professor and Chairman, Department of Biochemistry, University of California, Berkeley

Cornelius W. Pettinga, Executive Vice President, Eli Lilly and Company

Alexander Rich, Sedgwick Professor of Biophysics, Massachusetts Institute of Technology, *Cochairman*

Maxine Singer, Head, Nucleic Acid Enzymology Section, Laboratory of Biochemistry, National Cancer Institute

Robert L. Sinsheimer, Chairman, Division of Biology, California Institute of Technology

## PANEL FOR INQUIRY

Alexander A. Bayev, Institute of Molecular Biology, Academy of Sciences of the U.S.S.R., Moscow

Herbert W. Boyer, Professor of Biochemistry, University of California Medical School, San Francisco

Donald D. Brown, Director, Department of Embryology, Carnegie Institution of Washington

Liebe F. Cavalieri, Member, Sloan-Kettering Institute for Cancer Research; Professor of Biochemistry, Graduate School of Medicine, Cornell University

Stanley N. Cohen, Professor of Medicine, Stanford University Medical Center

Freeman J. Dyson, Professor of Physics, School of Natural Sciences, The Institute for Advanced Study

Daniel J. Hayes, Jr., Chairman, Cambridge Experimentation Review Board

Sheldon Krimsky, Associate Director, Program in Urban, Social and Environmental Policy, Tufts University

Bo G. Malmstrom, Professor of Biochemistry, Chalmers Institute of Technology, Göteborg, Sweden

Kurt Mislow, Hugh Stott Taylor Professor of Chemistry, Princeton University

Robert S. Morison, Class of 1949 Visiting Professor, Massachusetts Institute of Technology

Robert R. Murray, Jr., Professor of Pediatrics and Medicine; Chief, Division of Medical Genetics, College of Medicine, Howard University

Judith E. Randal, Science Correspondent, Washington Bureau, *New York Daily News*; Science Policy Editor, *Change* Magazine

DeWitt Stetten, Jr., Deputy Director for Science, National Institutes of Health

John Tooze, Executive Secretary, European Molecular Biology Organization, Heidelberg

George Wald, Higgins Professor of Biology, Harvard University

Milton R. Wessel, Attorney and Adjunct Professor, New York University School of Law

Luther S. Williams, Associate Professor of Biology, Department of Biological Sciences, Purdue University

Norton D. Zinder, Professor, Rockefeller University